Helmholtz

Helmholtz

From Enlightenment to Neuroscience

Michel Meulders

Translated and edited by Laurence Garey

The MIT Press
Cambridge, Massachusetts
London, England

© 2010 Massachusetts Institute of Technology

Helmholtz, des lumières aux neurosciences © ODILE JACOB 2001

MIT Press books may be purchased at special quantity discounts for business or sales promotional use. For information, please email special_sales@mitpress.mit.edu or write to Special Sales Department, The MIT Press, 55 Hayward Street, Cambridge, MA 02142.

This book was set in Syntax and Times Roman by Toppan Best-set Premedia Limited. Printed and bound in the United States of America.

Library of Congress Cataloging-in-Publication Data

Meulders, Michel.
[Helmholtz. English]
Helmholtz : from enlightenment to neuroscience / Michel Meulders ; translated and edited by Laurence Garey.
    p. cm.
Includes bibliographical references.
ISBN 978-0-262-01448-9 (hardcover : alk. paper) 1. Helmholtz, Hermann von, 1821–1894.
2. Scientists—Germany—Biography. I. Title.
Q143.H5M4813 2010
509.2—dc22
[B]
                                                                                        2009054153

10 9 8 7 6 5 4 3 2 1

Contents

Author's Acknowledgments

A book can never be written in isolation: The author may well give birth to an idea but needs the complicity of friends to realize the project. This book is no exception. So I wish to thank all those who have supported and helped me through its long gestation, both the original French text and this new English edition.

First of all, I thank my old friend Jean-Didier Vincent, who supported me from the beginning with the conviction that characterizes him, welcoming the idea of a work that, through the personality of an unusual scientist, aimed to shed light on one link in a chain, which was also a turning point in the history of neurophysiology and the philosophy of knowledge.

Next I have to thank Pierre Karli, through whom I met Claude Debru, director of the European Center for the History of Medicine at Strasbourg, and so encountered a group of scientists, philosophers, historians, mathematicians, and biologists that included, apart from Claude Debru himself, André Coret, Gerhard Heinzmann, Jacques Lambert, Charles Marx, Alexandre Métraux, and Philippe Nabonnand, all inspired to analyze works by German-language scientists, such as Hermann von Helmholtz, Ewald Hering, and Ernst Mach. I found in this group, in addition to sincere and generous friendship, an ideal context for the exchange of infinitely precise information and ideas. My gratitude can only hope to reflect what they brought me.

Among these friends I owe special mention to Alexandre Métraux of Mannheim University. For four years, this remarkable philosopher and historian of science followed the progress of my work, never failing to discuss my problems or criticize when he felt it necessary, enlightening my path with his vast knowledge and thoughts. However, should there be any errors in my work, I alone bear full responsibility.

I wish to thank Odile Jacob for admitting my project to the lineage of scientific works that she has given herself the task of publishing in the name of the progress of thought, as well as Jean-Luc Fidel, who was my editorial guide and helped me refine the presentation as a whole with his constructive suggestions.

Acknowledgments to the English Edition

I wish to thank my friend Laurence Garey most warmly for the magnificent work he has undertaken in translating *Helmholtz*. His task was arduous because the subjects in my book not only only concern physiology, which he knows well, but also more esoteric disciplines like philosophy and musicology. Thanks to his broad cultural background, Laurence has mastered the complexity of this multidisciplinary approach, which has revealed him as an outstanding interpreter. I should add that I have always been extremely impressed by his quest for perfection, as much in the explanation of difficult concepts as in the formal expression of his texts.

Laurence Garey was born and brought up in Peterborough, England. He was an undergraduate at Oxford, taking his BA degree in Animal Physiology in 1963. He then completed his medical degrees at St. Thomas' Hospital in London after a DPhil in Oxford as a student of Tom Powell. He has worked in brain research throughout his career in the Universities of Oxford, California at Berkeley, and Lausanne in Switzerland, Imperial College London, the National University of Singapore, and the UAE University (Al Ain, United Arab Emirates). His interests have spanned the structure and development of the mammalian visual system and the human cerebral cortex, including morphological correlates of schizophrenia. He has been active in the development of human brain banking, especially in relation to schizophrenia and multiple sclerosis. He was a local organizer of the first World Congress of the International Brain Research Organization (IBRO) in Lausanne in 1982, and he maintains his association with IBRO, having been chair of the Neuroscience Programmes Network and a member of IBRO's History of Neuroscience Committee. For many years, he was on the editorial board of *Experimental Brain Research*. He is also a professional anatomist, with a particular interest in applied and surgical anatomy, which has led him to travel widely teaching and examining, including membership of the Court of Examiners of the Royal College of Surgeons of England. He was a member of the Council of the Anatomical Society of Great Britain and

Ireland and is at present their archivist. He still teaches anatomy on an *ad hoc* basis when asked. Another of his interests is the history of neuroscience, and he has translated several works from French and German, including Brodmann's "Localisation in the Cerebral Cortex" and works by Jean-Pierre Changeux and Michel Jouvet, among others. He also indulges in another passion—the history of aviation.

This brief outline of Laurence's broad career helps explain why his translation often represents an improvement of my text, and I again express my profound gratitude to him.

Michel Meulders
Wavre, September 2009

Translator's Introduction

I first met Michel Meulders in November 1981 at a conference in Algiers, after which we traveled together exploring some of the treasures of the Algerian desert. I have remained in contact with him ever since. It was a great pleasure for me when in the summer of 2001, while he was staying in Switzerland, he presented me with a copy of his biography of Helmholtz, soon after its publication, and subsequently agreed for me to translate it.

Michel has a long and outstanding career in neuroscience, and he lived the profound changes in neurophysiology of the 1960s.

Born and brought up in Belgium, he was a medical student at the Catholic University of Louvain. During his studies, in 1950, being one of the top students, he was offered the chance to undertake some research in the Laboratory of Biophysics of the Institute of Physiology under the supervision of Jean Colle. He joined two other students, Jan Gybels and Jean Massion, and together they worked on several projects, including neuropharmacology of frog nerve, experimental epilepsy and respiratory reflexes, and their relationship to postural tonus and the cerebellum. The whole of the physiology of the nervous system at that time was indelibly marked with the influence of Charles Sherrington, and the Louvain team was no exception to the rule.

After qualification in medicine, he was appointed as a junior clinical neurologist under Ludo van Bogaert in Antwerp. During this time, he learned from his contact with patients that the basic medical sciences, like physiology and anatomy, must be seen in the context of the whole person and forged a link between those sciences and psychology, which was to be a hallmark of his approach thereafter.

It is notable that the head of the Institute of Physiology was Joseph Bouckaert, a scientific descendent of Ernst Brücke, and therefore of Johannes Müller, both of whom play major roles in the biography of

Hermann Helmholtz, as we shall see. Michel was also influenced by Albert Michotte, a student of the psychologist Wilhelm Wundt, himself Helmholtz's assistant.

From 1958 to 1959, he worked in the laboratory of Giuseppe Moruzzi in Pisa, where exciting work on the states of vigilance and their relationship to the newly discovered reticular formation was in progress, another formative experience for the young psychophysiologist.

He returned to Louvain and worked for another two years with Jean Massion, in particular on the electrophysiology of the thalamus. With the latter's departure to make a notable career in France, Michel was the last of the three original medical research students, and he remained true to Louvain ever since. He soon took up the chair of Psychophysiology in the Institute of Psychology, where he further developed his ideas on the need to study the whole person in both his research and teaching.

By then the former Laboratory of Biophysics had become that of Neurophysiology, before the university underwent a disruptive and traumatic split along linguistic lines that resulted in the formation of Flemish and French language campuses and the redeployment of personal rather arbitrarily between the two. Michel became head of the new Francophone Laboratory of Neurophysiology. As the concept of "neuroscience" was emerging, his team worked on a variety of themes, including the function of the motor system and manual dexterity, the effect of vigilance on properties of visual neurons in the pulvinar of the thalamus, eye movements and gaze in relation to the superior colliculus of the midbrain, as well as hypothalamic mechanisms.

After 18 years in this position, Michel was becoming more and more involved in university administration. He stepped down from the headship, becoming Dean of the Medical School from 1974 to 1979 and member of the Rectorate from 1983. He was president of the Research Council of his university from 1980 and became Prorector in 1991. He is also an enthusiastic member of the Belgian Royal Academy of Medicine and was its president in 1986. He became an Associate Member of the French National Academy of Medicine and President of the Helmholtz Academy in Strasbourg.

In addition to his mastery of neuroscience, Michel is an accomplished musician, a respected violinist, and cofounder of the Symphony Orchestra of his university.

He is now Emeritus Professor of Neuroscience and Honorary Prorector of the Catholic University of Louvain. He continues to follow the neurosciences, of which he might qualify as a "founder member," and he continues to broaden and deepen his interest in the history of the subject with, since Helmholtz, works on Ernst Mach and William James, of whom he has a biographical study forthcoming at this very moment.

We must thank Marco Piccolino, translator of the Italian edition of Helmholtz, for allowing us to use some of the plates from his book.

Laurence Garey
Perroy, September 2009

Preface

Although he remains one of the most remarkable figures of German physiology of the nineteenth century, Hermann von Helmholtz has curiously remained little known outside German-speaking communities. This paradox is doubtless partly due to the language barrier but also to the difficulty of confronting his research, rooted in the complex worlds of philosophy, physics, and the arts, notably music. Furthermore, Helmholtz relied to a great extent on psychology as unshakably associated with physiology, a reliance that he judged necessary for understanding major psychophysiological functions, such as vision and perception of space. This association had difficulty surviving the developments in physiology that marked the first half of the twentieth century, dominated by Charles Sherrington,[1] who diverged radically from Helmholtz's concepts.

Sherrington's model of the organization of the nervous system added much to the progress of neurophysiology, but it relied essentially on a hierarchy of reflexes[2] and was hardly suited to global psychophysiological problems such as sleeping and waking, voluntary movement, sensory perception, emotional behavior, or the mechanisms of motivation. Indeed it again became necessary in the second half of the twentieth century to take more notice of psychological data and adopt a methodology that allowed a global approach to these problems without submitting to the temptation to entirely reduce psychology to physiology.

Through a singular irony of history, it was especially researchers in psychology (e.g., Richard Gregory) who rescued Helmholtz, so famous in the nineteenth century, from the relative oblivion into which the school of Sherrington had plunged him.[3] Progressively more research was undertaken in which data derived from both physiology and psychology were studied conjointly, thus throwing a more global light on nervous function. This experimental approach, initiated by Helmholtz, was one of the principal sources of what has today become neuroscience.

This is why it seemed to me useful to study this pioneer more thoroughly and to try and make this exceptional scientific figure better known to the general public. That is the aim of this book. Nevertheless, after reading the monumental biography of Helmholtz by his pupil Leo Koenigsberger,[4] I realized that the complexity of my project far exceeded what I had imagined. I had not only to decipher a biography that was certainly remarkable but that often seemed more akin to hagiography than an objective, critical study. In addition, it was necessary to analyze the essential works of Helmholtz in their original language and situate them in their historical context in order to grasp their deepest meaning. But what I was far from suspecting when I started my work was the central role in the genesis of his scientific work played by great philosophers such as Emmanuel Kant,[5] but also idealists such as Johann Gottlieb Fichte,[6] and above all the *Naturphilosophen*, so specific to German thought that it is doubtful if the literal translation of the concept as "natural philosophy" is adequate. However, let us agree to use the term *natural philosophers*.

Helmholtz was trained by one of the best physiologists of his time, Johannes Müller,[7] and resolved to use only a materialistic, empirical scientific method in his research, free of metaphysical constraints, based on experimental data and the results of their mathematical analysis. He was thus in the direct line of Kant but also of the English empirical philosophers. However, he was directly opposed to the idealist and natural philosophers who interpreted nature from a viewpoint based on presuppositions that were inaccessible to experimentation. In the quarrel over vitalism, for example, he intervened vigorously against any attempt to explain the nature of life by metaphysics, and he also opposed some of the generally poorly known scientific ideas of Goethe who had written a romantic theory about colors.

To help my readers and better situate my project, it seemed useful to first briefly sketch the atmosphere and cultural ambiance of the 1820s in Germany, the time when our physiologist was born. I have tried to do this first by means of a short fictional story. While strolling in the park of Sanssouci at Potsdam, Helmholtz's father expresses some rather disillusioned ideas on the end of the Age of Enlightenment. Then Goethe, close to death but still romantic, reflects bitterly on the incomprehension encountered by his scientific work. Both sense the end of an era and apprehend the one that is beginning.

After a first chapter describing Helmholtz's youth in his familial and cultural context up to his entry to the University of Berlin, the second chapter is dedicated to natural philosophy because of its importance for the comprehension of science and medicine in Germany at the beginning of the nineteenth century and because it is relatively little known outside Germany. In the subsequent chapters, Helmholtz's principal scientific contributions are summarized and analyzed, accompanied by a description of his cultural background and a few biographical details that are indispensable to situate ourselves and humanize a rather secretive personality. These details also permit us to understand how this personality so ably programmed and accomplished his career as both scientist and statesman. Particular attention is paid to his crucial discussion of the speed of conduction of the nerve impulse and his invention of the ophthalmoscope. Considerable space is devoted to his research on color vision and his opposition to Goethe's scientific works, which we examine. They have been discredited but are nevertheless rather attractive and far from devoid of significance for the advancement of our knowledge.

Helmholtz published two works of major importance. First is his *Handbook of Physiological Optics*,[8] in which he concentrated on his research on vision and visual perception. This work contained the essentials of his physiological doctrine and threw light on the way in which he conceived the complex relationships of physiology with philosophy and logic. Then his *Sensations of Tone as a Physiological Basis for the Theory of Music*,[9] which was one of the jewels of the scientific literature of the nineteenth century by the richness and imagination of the experiments that he described but also by the passion that inspired the author in his utopian desire to explain the beauty of music.

My work ends with an analysis of a speech by Helmholtz, a few months before his death, in honor of Goethe, in which he recognized that an artist worked at a different level of consciousness from a scientist and that his artistic intuition revealed truths about the human mind that remained inaccessible to scientific research. In this sort of testament, Helmholtz affirmed that the scientist and artist were capable of acquiring new knowledge on the condition that each remained in his own methodological domain. With these words, Helmholtz established an eminent place among the philosophers of science of the nineteenth century but also of today.

Prelude

I think, therefore I am.
—Rene Descartes[1,2]

I am, therefore I think.
—"Cogito of the Enlightenment"[3]

Light and spirit, the one reigning in the physical world the other in the moral, are the highest imaginable indivisible energies.
—Johann Wolfgang Goethe[4]

Dawn at Sanssouci

After passing through the majestic Brandenburg Gate at Potsdam, the young teacher from the Potsdam *Gymnasium*, August Ferdinand Helmholtz,[5] headed toward the palace of Sanssouci at a slow, measured pace. This stroll was a habit for him; he liked to reflect on philosophy, and at these moments he felt intensely free, freer even than with his friends at home, where he did not always dare express his deepest thoughts in view of how overwhelmingly suspicious the Prussian authorities were toward intellectuals. It was the morning of August 31, 1821. His wife, Caroline Penne, daughter of an artillery officer and remote descendant of William Penn, founder of Pennsylvania in 1681, was expecting their first child in a few days. "What would the future hold for this infant?" wondered the young teacher.

At a turn in an alley, he glimpsed the palace of Friedrich II,[6] its ochre facade already glowing brightly in the rays of the rising sun that had not yet dispersed the mist that covered the surrounding countryside. He could not avoid thinking of the engraving by Daniel Chodowieki of characters in the shadows of a dark forest heading toward a building, like a palace, already lit by the sun's rays, while a veil of mist still

Figure P.1
Aufklärung: engraving by Daniel Chodowiecki.

extended behind it (figure P.1). The artist made the following comment on the engraving, titled *Enlightenment*:

This ultimate work of Reason ... has until now no generally understood allegorical symbol ... except the rising sun. Doubtless this symbol will remain for long the most pertinent, because of the mists that will always rise from marshes, from censers and from victims burned on the altars of idolaters, that can so easily veil it. But if the sun has risen, the mists can do no further harm.[7]

The palace that August Ferdinand saw before him was for Friedrich II, a privileged place from where the light of reason would radiate over Prussia and Europe. It was at the top of a hill surrounded by six semi-circular terraces of alternating bay windows and trellised vines. Behind the upper terrace, in the screen of the vineyards, the palace stretched out, its single-story facade decorated with enormous caryatids, bacchants, and maenads sculpted in the sandstone. In the center of the facade was a curved pavilion with an almost provocative inscription: *Sans Souci*.[8] He knew that on the other side was the official facade, almost severe, with its *cour d'honneur* enclosed by a semicircle of double Corinthian colonnades. This superb edifice was the work of the architect Georg Wenzeslaus von Knobelsdorff,[9] a friend of Friedrich who wanted to build a place reserved for the arts and private meetings with his friends. The "philosopher of *Sanssouci*," as Friedrich like to call himself, had installed a magnificent library as well as a rococo music room, with white and gold walls, decorated with enormous mirrors, where he often played the flute, accompanied on the harpsichord by Carl Philipp Emanuel Bach. Indeed, it was at Potsdam, at the royal residence in the town itself, that he proposed to Carl's father, Johann Sebastian, when he visited him in 1747, a musical theme of his own composition that his visitor had immediately, and in his presence, developed into thirteen superb variations known under the name of *Musical Offering*.

It was a real honor to be invited to the king's table, and Voltaire, Friedrich's guest from 1750 to 1753, described in his memoirs some of these renowned meals:

One dined in a small room in which the most remarkable decoration was a painting created by Pesne, his artist, one of our best colorists. It was a fine priapic scene. One could see young men kissing girls, nymphs beneath satyrs, cupids playing at Encolpe and Giton ... Meals were rarely less philosophical. A spectator listening to us and seeing this painting would have thought he was hearing the seven sages of ancient Greece at the brothel. Never, in any place in the world, would one have spoken with such freedom of all man's superstitions, and never would they have been treated with more jesting and scorn. God was respected, but all those who had deceived mankind in his name were not spared.[10]

August Ferdinand Helmholtz respected religion and the values it was deemed to help made known. Above all, however, he was committed to liberty. In 1811, at age nineteen, he enrolled at the Faculty of Theology of the University of Berlin, but he was disappointed by the excessive orthodoxy he discovered there. After the heroic events of the battle of Dresden in 1813, in which he had participated as a volunteer against Napoleon's army, he left the faculty and turned his studies toward philosophy, which he much enjoyed, and philology, finally being appointed as a school master in Potsdam. The young teacher made his presence felt later, in 1845, by a violent diatribe against the integrism of the Evangelical Church, which considered as heretics and politically suspicious all who did not follow its precepts, which it considered to be as infallible as those of the pope.[11]

Everyone had dreamed of this liberty since the beginning of the Enlightenment. It was naturally linked to the emergence of rationality and its powerful movement for intellectual emancipation. That aim was far from being achieved at that time. Furthermore, during the eighteenth century, much water had flowed under the bridge of ideology. The philosophy of the Enlightenment, the emancipated daughter of Cartesianism, owed to Descartes—and to Malebranche—a taste for reasoning, a search for intellectual evidence, and, above all, the presumption to exercise one's judgment freely and always show a spirit of methodical doubt. However, since Descartes, the sources of knowledge had become radically different (see chapter 2, this volume). "I am, therefore I think" had in a way become the *cogito* of the Enlightenment, very close to the Cartesian *cogito*. Very close, but with a diametrically opposite meaning. As d'Alembert said, "The arms that we use to fight him (*Descartes*) are no less his because we turn them against him."[12]

Certainly, Friedrich II played an emblematic role in this century of Enlightenment, but the progressivism that he paraded was ambiguous. He swelled the ranks of his army thanks to the recruitment of numerous foreigners, thus allowing him to prove his freedom of spirit. He tolerated two religions in Prussia but gave the best appointments to Protestants. He protected Kant at Königsberg, but other renowned writers or thinkers took care not to reside in Prussia, which Gotthold Ephraim Lessing described as the most enslaved country in Europe.[13]

Plunged in these reflections and a little perplexed, August Ferdinand continued his stroll past the mythical palace. The sun's heat was still tempered by a light breeze that had arisen; it came from the west, and it made him think of France and the British Isles. He thought with a certain emotion about the Encyclopedists, whose confirmed atheism so displeased Friedrich, who had obstinately strived to realize a synthesis of contemporary knowledge and so built the basis of a civilization of liberty and progress.

There were also the empirical philosophers, British for the most part. First was John Locke,[14] who opposed Cartesian metaphysics and who had the effrontery to affirm that our ideas stemmed exclusively from two sources: sensation and reflection. Then there was David Hume,[15] who, in the name of everyday experience, rejected all notion of causality to retain only that of a more or less frequent associative relationship between phenomena. This obviously excluded all reasoning based on causality, a dead end in the philosophy of rationality that Kant was to avoid later.

Depite the turmoil of his reflections, our stroller remained calm because Friedrich had planned ahead for his subjects. He raised his eyes and saw that he was close to the mill next to the palace of *Sanssouci*. He recalled the story of the miller from whom Friedrich wanted to buy it as he was disturbed by the noise of the sails during the night. The miller refused the king's offer; when the king began to threaten him, he replied, "Sire, there are still judges in Berlin." The king gave in, in recognition of the limits of his power over his people.

Something else that reassured August Ferdinand about the importance of the Enlightenment was the famous reply that Kant gave in 1784 to the question of what Enlightenment was. "Enlightenment is man's release from his self-incurred tutelage. Tutelage is man's inability to make use of his understanding without direction from another. Self-incurred is this tutelage when its cause lies not in lack of reason but in lack of resolution and courage to use it without direction from another. *Sapere aude! Have courage to use your own reason*! That is the motto of Enlightenment."[16]

Continuing his stroll and his day dreams, the teacher came down the hill and walked toward the park. It still had its central alleyway that stretched out parallel to the palace from the obelisk at the entrance to the enormous *fanfaronnade*, as Friedrich himself called it, which formed the new palace.[17] This baroque edifice, built in 1763 at the end of the Seven Years War, had been conceived as a symbol of Prussian power. Friedrich did not enjoy living there, and it became a mere summer residence for members of the royal family. The monotony of the straight alley was broken by several circles decorated with fountains and statues. He decided to rest a moment near the lovely goddesses of the arts and sciences, listening to the rustle of the leaves and the splashing of the fountains.

Why then, he thought, was the German spirit satisfied in exploring the certainties that the Enlightenment promised? Why was the German spirit so sophisticated that it plunged body and soul into the frenzy of *Sturm und Drang*? Especially during the lifetime of Friedrich II, the

very heart of the eighteenth century? Why the return of today's philosophers, such as Friedrich von Schelling,[18] Georg Hegel,[19] and Arthur
Schopenhauer,[20] to the old demons of the irrational? Why did Goethe
reject Isaac Newton's scientific discoveries in favor of a romantic view
of nature?

A few hours later on August 31, 1821, the first son of August Ferdinand Helmholtz and Caroline Penne was born: He was named Hermann
Ludwig Ferdinand and was baptized in the Lutheran faith.

Twilight at Weimar

At the very moment that August Ferdinand was deploring the decline
of the Enlightenment in which he believed so ardently, and was fearing
that his country would finally abandon this so promising pathway of
reason and science, the illustrious poet Johann Wolfgang Goethe[21]
was deeply saddened for very different, but equally bitter, reasons. He
was famous throughout Europe for his enormous poetic and prosaic
talent, as well as for his dramatic works from *Götz von Berlichingen* to
Faust. He was received by Napoleon in person, who had read his *Werther*
seven times during his campaign in Egypt and had proclaimed to his
generals, "This is a man!" Nevertheless he felt deeply wounded because
of the incomprehension that his scientific works inspired among his
contemporaries.

He was now 72 years old. Privy councilor and minister of the Grand
Duke Karl August of Saxe-Weimar, Goethe had transformed Weimar,
a comic-opera capital of 4,000 inhabitants situated south of Potsdam,
into a cultural hub that was sometimes called the Athens of Germany.
Johann Gottfried von Herder,[22] Christoph Wieland,[23] and, especially,
Friedrich von Schiller[24] had been his close friends, but they were all dead.
In addition to his administrative and literary activities, he had carried
out scientific research in morphology, botany, mineralogy, and even
optics. Despite his efforts, he felt very misunderstood. People preferred
official science, which attempted to explain nature by reason, with the
help of experiments, physical analysis, and mathematical quantification,
whereas his own approach was almost the opposite and was considered
as romantic and meta-rational.

In his study from dawn, his first thought was generally for Faust, his
inseparable companion, whose states of mind and revolts, but also whose
hopes and passionate dialog with the forces of evil, had always reflected
his own anguish with regard to knowledge. Had he not, in the first lines

of his primitive *Urfaust*, written in 1773 when he was only 24 years old, expressed all the vanity of bookish and compartmentalized science?

Alas! Philosophy,
Medicine and law
And sadly too theology,
I studied you in full with ardent effort.
Now, here I stand, poor Thor,
No wiser than before.
...
And see that we can know nothing

And how bitter he must have been to make the spirit of the Earth, whom Faust invoked, reply:

You are the equal of the spirit that you conceive,
Not of me!

He meant that human knowledge had limits that no one could transgress.

In 1769, while still young, he had made a brief and disappointing incursion into the world of alchemy and the occult by reading Paracelsus and especially Swedenborg, the visionary mystic from "the shadows of legend."[25] Despite passing periods of despair, his desire to improve his knowledge had always been foremost, and he was convinced that his scientific legacy to posterity was more important than the totality of his literary works, even more so because he believed he had produced a true methodological charter for research.

This relied on a double conviction. First, nature was more than the sum of its parts: It constituted a coherent whole animated by an "essence" capable of acting on those parts. Second, sensory perception was the only tool that permitted us to separate, distinguish, and then reunite phenomena that attracted our attention to finally establish something that appeared to us as an entity and gave us more or less satisfaction.[26] Furthermore, he was convinced that nature was visible to whoever looked for it: "Nature has no secret that it does not unveil somewhere before the eyes of an attentive observer."[27]

He remembered having sent to Schiller, who had been his greatest friend in Weimar and with whom he often discussed his research, a summary of his point of view on various types of phenomena.

An *empirical* phenomenon, which all men perceived in nature, could be raised to the rank of a *scientific* phenomenon by experiments presenting it in different circumstances and conditions than those in which one knew it first. Finally came the *pure* phenomenon, the result of what had been perceived and the experiments.

He introduced this pure or *primeval* phenomenon (*Urphänomen*) in his *Theory of Colors*.[28] In the end, everything was subject to "higher rules and laws not revealed to our intelligence by words and hypotheses, but to our intuition by phenomena. We call them primeval phenomena for nothing appears to be above them." Concerning a primeval phenomenon, it could be that "one refuses to recognize it as such, that we seek something else behind it and above it, whereas we should admit that there lies the limit of our perception."[29]

He added later that when the primeval phenomenon had manifested itself, one must stop, be reassured, and be satisfied. Goethe's concept of primeval phenomena is complex. It has been widely studied, notably by Lacoste,[30] who said, "the primeval phenomenon ... reveals itself to the intuition when the mind has grasped enough examples after many 'experiments'. Its unity appears little by little, but it does not give access to a higher reality, to an intelligible world. In this sense it is at once the aim and the limit of knowledge." He also showed how the term was used later in other contexts. For example, Cassirer, in his works of the 1920s, saw a primeval phenomenon as the general function of expression that permitted the constitution of different symbolic forms such as myth, language, science, and art.[31]

Goethe remembered that well before he formalized the primeval phenomenon, he had found it necessary to use the concept in botany and especially in morphology. To study dissected elements of an organic body separately and then revive the destroyed creature and consider it as alive and healthy again was the work of a physiologist. So, if physiology was the intellectual process by which we dealt with both the living and the dead, by examination and by deduction, and of which the manifestations and effects always remained mysteries, it was easy to understand why physiology would perhaps always remain in the background. Man always knew his limits but rarely wanted to admit them.[32] In the end, physiology, as a science that might reveal the key to life, was an impossibility.[33]

Suddenly the figure of Isaac Newton[34] came to his mind, which made him very bad tempered. All his life he had fought the delirious theories of the Cambridge physician, who had not hesitated to affirm that white light could be decomposed by passing it though a prism into a beam containing the colors of the rainbow. Even worse, one could then make these different colors reconverge through a set of lenses and thus recreate the original white light. Moreover, he was angry with Voltaire who had been one of the principal proponents of the vulgarization of this theory and had succeeded in imposing the dogma of the Newtonian gospel on intellectual Europe. For he, Goethe, had discovered that the couple light-darkness was a primeval phenomenon, and that it was pointless to try to dissect through the physical intermediary of a prism that which was at the limit of human perception: "May the scientist leave primeval phenomena in eternal peace and splendor."[35] Even Faust gave him support[36] when, while awaiting dawn, he saw the sun gradually rise, which obliged him to shade his eyes and turn away from the light, exclaiming "May the sun keep away!" when he saw the superb colors of reawakening nature appear.

Unfortunately, the scientific world had been hard on Goethe and deemed his theory of colors romantic, poetic, and absolutely not prosaic, explaining well-known phenomena with the artificial language of transcendentalism. He had been deeply saddened by this but did not lose his fighting spirit, persisting in his efforts to have his views accepted, going as far as to write to one of his correspondents that in 1821 the Newtonians found themselves in the situation of Prussia in October 1806: They still believed they were winning the tactical battle, whereas they had already been defeated strategically long before. This alluded to Napoleon's war against Prussia in 1806, which ended tragically at Jena.[37] As he was getting older, it was even more important for him to fight for his scientific cause, as well as for his image as poet and playwright, novelist, and essayist, which brought him into daily contact with the great men of this world, who jostled for the privilege of a conversation with him.

Without knowing it, August Ferdinand Helmholtz and Goethe agreed about one thing: Both were living the end of their respective worlds. One had seen the blossoming of the rationality of the Enlightenment, and the other has witnessed the awakening of the romantic culture. Those first twenty years of the nineteenth century certainly did not totally reject the heritage of the eighteenth, but the profound political and social transformations in Germany saw the birth of often radical metamorphoses in moribund cultures. Among these metamorphoses, Hermann, August Ferdinand's newborn son, would be spokesman for a self-proclaimed empirical and rational science. But numerous German philosophers, idealists, and metaphysicians resolutely rejected science that was worthy of the name.

1   Helmholtz: From Potsdam to the Pépinière

Iητροσ φιλοσοφοσ σοθεοσ

[The physician who is a philosopher is equal to the gods.]
—Hippocrates

Die Gedanken stehen in demselben Verhaeltnis zu dem Gehirn, wie die Galle zur Leber oder der Urin zu den Nieren.

[Thought is related to the brain as bile to the liver or urine to the kidney.]
—Karl Vogt, physiologist

To which of these citations, neither of which can claim to be devoid of arrogance or conceit, would Hermann Helmholtz, whose career and work we shall now pursue, have rallied? Doubtless to the second because, as one of his closest friends Emil du Bois-Reymond reported, he considered one of the main principles of natural science to be that nature, including perception, thought, and free will, should be explicable, without which all research in the field would be senseless.[1]

As to the citation by Hippocrates, he considered it as the standard of old deductive medicine, against which he had always reacted forcibly. But that did not stop him, in a famous speech in Berlin in 1877, from adding a very subtle comment. Citing Hippocrates that "the physician who is a philosopher is equal to the gods," he went on: "We could accept it if we could determine the meaning of the word philosopher. For the ancients, philosophy still included all theoretical knowledge; their philosophers placed mathematics, physics, astronomy and natural history in close relationship with their actual philosophical and metaphysical considerations. If you mean by Hippocrates' physician-philosopher a man with complete insight into causal relationships between the processes of nature, then we could indeed say, like him, that such a man can help us like a god …

the ideal that our science should strive for. Whether it will ever succeed, who can really say?"[2]

All the complexity, but also the prudence of expression, of Helmholtz's philosophy is contained in these lines. A confirmed materialist in his scientific approach, he nevertheless had a sufficiently broad intellectual and cultural outlook to be able to estimate just how far he had to go and to open his mind to humanistic and idealistic questions.

The Potsdam Years

But all that was far removed from little Hermann, born in Potsdam in 1821, whom we meet at his childhood home, a large, spacious middle-class house on a main road in the center of the town, which one can still see today. At the time, Potsdam had a little more than 20,000 inhabitants, half of whom were in the military or who depended on the army. The royal household, officers drawn almost entirely from the nobility, and members of the Prussian higher administration all contributed to making Potsdam a one-class society. The Helmholtz family, although belonging to the intellectual middle class (*Bildungsbürgertum*[3]), nevertheless had access to high society because their father was a high school teacher and a veteran of the war of independence, and their mother was the daughter of an artillery officer. The list of 23 godparents gathered for the baptism of the new baby featured top-town administrators, as well as an uncle physician-general, who later put all his energy into facilitating the young man's entry to the military school of medicine.

When he was in his 70s, von Helmholtz, raised to the peerage in 1883, recalled that in his youth he was often sick and had to stay in bed. His parents looked after him well; they showed him picture books and gave him wooden toys to build, which, he believed, helped him learn very early about geometric relationships between objects in space. He learned to read quite early and was quickly able to widen his field of activity, but he soon realized that memorizing was a real problem for him, especially concerning basic facts such as left from right or, later, foreign vocabulary or texts to recite by heart. In old age, these troubles of memory became a real ordeal for him.

When he grew up, he was constantly plunged into the books of his father's library or, together with a friend, putting together small optical instruments from old spectacles and a botanical magnifying glass, or experimenting with elementary chemistry by mixing acids. In addition he loved walking in the Potsdam countryside, a taste that he indulged all his life by an annual walking holiday in the Alps lasting several weeks, an opportunity to pursue his studies on the formation of glaciers, which have become classics.[4]

The cultural life in Potsdam was very active, and the Helmholtz family made the most of it. They frequented people who were interested in the theater and the arts in general. Father Helmholtz enjoyed and indulged in painting. Little Hermann had learned the piano and liked it very much, which proved most useful later when he wrote his treatise on the physiology of music.[5]

Among the many people he met at home was his father's friend Immanuel Hartmann Fichte, a philosophy master at Tübingen and son of the famous Johann Gottlieb Fichte, the philosopher and former rector of the University of Berlin, who had been the two boys' professor when they were young. Johann had had a great influence on August Ferdinand, who much admired him and emphasized this to Hermann in a moving letter sent in 1857, two years before his death. In this letter, he asked his son not to underestimate the old philosopher's opinion on anthropology and entreated him to always take account, in his effort to understand nature through observation, of knowledge derived from philosophy and vice versa. Schelling and Hegel, and their pupils, had committed the error of neglecting the results of observation and had wrongly believed they could "construct" the whole world in this way, an error that neither Fichte father nor son committed. "Self knowledge," he added, "is that toward which a divine instinct pushes man; but it is not possible without knowledge of nature, so natural scientists must accept that in searching to comprehend the philosophical aspect of their results, however insufficient they may be at that moment, they seek to respond to the highest needs of the soul."[6]

Discussions on philosophy were common in cultivated society in Germany and woven into them was a complex religious thread. Without it, philosophical reflection, whether concerning the Enlightenment, natural philosophy, or German idealism, would not have developed as it did. Protestant churches formed the state religion through much of Germany, especially in the north and the center in Prussia and Saxony, while Catholicism dominated the southwest in Bavaria and the Rhineland. Indeed, it was in the north and center that the major German cultural and philosophical movements saw the light of day, from Leibniz to Kant and Schelling, from Lessing to Goethe.

Perhaps we can find a reason, at least partial, for this selective cultural fertility in the very structure of the Protestant religions. In these offspring of Luther and Calvin, contact between the faithful and God was made directly, without intermediaries, and the only moral authority was the pastor, who was not subordinate to a bishop or a pope. This conferred a personal responsibility and an undisputed freedom of mind which explained that Protestant theologians were often intellectuals of the highest stature. But Protestantism, so favorable to the blossoming of individual potential, also suffered crises, notably the pietist movement. Provoked by the rigid orthodox dogmatism of the official religions, pietism imposed itself progressively from the beginning of the eighteenth century. It was animated by an often very affective desire for a deepening of religious life and a great spirit of tolerance. It was very influential in Prussia, where it encouraged individualism and a national ideal. Kant, Goethe, and Schelling were born into families steeped in pietism. Wanting to make German culture better known to the French, Heinrich Heine wrote in 1832: "I have mentioned freedom of mind and Protestantism together; in fact in Germany there exists a friendly relationship between them. ... (Protestantism) has allowed free research into the Christian religion and freed minds from the yoke of authority. Free research has been able to spread roots throughout Germany and science has been able to develop independently."[7]

This conclusion was perhaps too unequivocal if we consider the frequent migrations of German philosophers from their intolerant state to a neighboring more tolerant one, and if we recall, for example, the precautions that Kant took to avoid the susceptibilities of Prussian censure. Heine's statement was nevertheless valid because it expressed his conviction that Protestantism was a more fertile ground than most for the explosive expansion of the philosophical movements of the Enlightenment and the romantic period.

When his father's mathematician friends met at their home, Hermann was present at their discussions and heard for the first time talk of *perpetuum mobile*, perpetual motion, and the numerous unsuccessful attempts that had been made to demonstrate its feasibility. According to Koenigsberger, that was when he decided to reflect on this problem himself, completely independently (see chapter 5).

At his high school, he followed the classical curriculum and was a very good student without being exceptional. Physics and mathematics fascinated him; he took Latin and Greek, but also French, English, Italian, Hebrew, and even the rudiments of Arabic. He was clearly not attracted to reading Cicero or Virgil because at such times he was sometimes caught in the middle of complex calculations or measurements of the path of light rays on a drawing of a telescope destined for his own amateur experiments in optics. In his senior classes, he had his father as teacher of physics and mathematics, as well as of philosophy and German. As such, the latter had to give his opinion on his son's school-leaving essay on *Ideas and Art in Nathan the Wise* by Lessing. His father expressed his complete satisfaction about the formal aspects of the essay but was rather critical about its substance: For him it showed that Hermann's comprehension was better than his reflection because the central themes of the text were not sufficiently analyzed, thus weakening the whole.

It was obvious that young Hermann should choose a career in physics and mathematics, in which he excelled. Unfortunately, the financial circumstances of his parents did not permit him to envisage such a risky career. Despite his great desire to embark on such studies, he had to accept to consider medicine, which he could study free in the context of the army and which opened better socioeconomic perspectives for the future. Encouraged, and doubtless supported, by his mother's uncle, physician-general Mursinna, he took the entrance examination for the Friedrich-Wilhelm Medicosurgical Institute, formerly called the *Pépinière*. He passed his examination very successfully, including a final essay on *The Contribution of the Study of History to the Scientific Education of the Mind*. On September 26, 1838, he left Potsdam for Berlin with his books and his piano, on his way to his university studies.

The Pépinière

The Friedrich-Wilhelm Medicosurgical Institute had only earned this pompous title in 1818, when the wars for independence had provoked the development of Prussian nationalism. Everyone still called it the Pépinière twenty years later. It was situated between the Charité hospital and Humboldt University, and the medical students were subject to strict discipline and hard work. Part of their teaching, particularly of human sciences, was on site, and the rest was done at the hospital by the medical faculty of the university. At the end of their course, students of the Pépinière had the same medical qualification as students from the faculty, but the rigidity of their curriculum was more like that of a technical school than a university as imagined by Wilhelm von Humboldt with his concepts of freedom to learn and to teach (*Lernfreiheit, Lehrfreiheit*).[8]

Although very busy with his work, to which he consecrated much energy despite his fragile health—he had frequent fainting attacks and migraines—Hermann nevertheless enjoyed a considerable social and cultural life. He played a lot of music in his room and also at the home of his parents' friends who invited him each Sunday. In his letters to them, he described evenings at the theater or the opera, where he had seen performances of *Hamlet* and *Faust, Don Juan, Eurianthe*, and *Iphigenia*. When he returned to his room, he immersed himself in relentless reading, especially Homer and Kant, but also *Faust Part Two* and Byron.[9]

Among his teachers was one whom he particularly admired: Johannes Müller.[10] Müller had noticed the intelligent young medical student and had allowed him to visit his own research laboratory. It was unusually small for a professor of his stature and scientific prestige, especially compared with those in Vienna or Munich. However, it was a very attractive place, thanks to the personality of its head, and exceptional scientists were trained there who were later to form one of the most remarkable physiological networks of the nineteenth century. Hermann received a good-willed welcome from two of the older assistants already working there, Emil du Bois-Reymond[11] and Ernst Wilhelm von Brücke.[12] Later, one of Brücke's most illustrious students, in Vienna from 1876 to 1882, was Sigmund Freud,[13] who all but worshipped him. Soon the initial good will gave way to friendship between the three men, which, as we shall see later, became a focus of creativity and mutual esteem that would last their whole lives. Helmholtz was the youngest of the group, but he had a certain superiority over his friends due to his excellent knowledge of mathematics and physics.

It is very true: what is below is like what is on high, and what is on high is like what is below, to perform miracles of one thing.
—"The Emerald Tablet" of Hermes Trismegistus, mythical Egyptian alchemist[1]

As Helmholtz began his studies around 1840, medical teaching was nearing the end of a long reform in which the traditional curriculum had, according to him, gradually given way to a new scientific spirit that rejected tradition and insisted rather on a basis of personal experience. The ideas of John Brown[2] and Albrecht von Haller,[3] as also of vitalism (see chapter 4), were on the decline, and the excesses of the deductive methods of ancient medicine were no more than a dogmatic ruin.[4] But what was really the situation? Romanticism and natural philosophy had profoundly impregnated the theory and practice of medicine between 1790 and 1815, and that influence was still palpable at the time of Helmholtz. Furthermore, Kant's philosophy was enjoying new prestige among German physicians and scientists because it seemed to them to defend a new scientific spirit.

Precursors of Natural Philosophy: Bruno, Spinoza, and Leibniz

In general, natural philosophy was *monist* and *immanentist*, which meant that the mind and body constituted a single reality, and that God was within the universe, not outside it. It was therefore radically opposed to Descartes, for example, who was doubly dualist, separating the soul from the body-automaton and putting God outside nature, which depended on his sovereignty for the laws by which it functioned. It also opposed Galileo and Newton because they reduced nature to an abstract and rigid geometrical and mathematical framework. At the heart of the debate was the relation between the infinite and the finite, between God and nature, between mind and body. This debate stretched back to the philosophers of antiquity. Three great philosophers must be mentioned because of their importance for the emergence of natural philosophy: Giordano Bruno,[5] Baruch Spinoza,[6] and Gottfried Wilhelm Leibniz.[7]

A visionary philosopher, defender of Copernican heliocentrism, and totally independent spirit, Bruno affirmed forcibly that the intellect could overcome the limitations of our finite senses: "I render the heavens and plant myself in infinity,"[8] he said, comparing his mind to the flight of a javelin. Tto ensure a global coherence between the world of science and that of philosophy, Bruno had the audacity to propose a unified vision of an immanent God in an infinite cosmos. He affirmed that nature itself, more than the mind that measured it, was the real divine power and the order imprinted in all things.[9] He was almost forgotten by later generations, and it seems that Spinoza, whose ideas were so close to those of Bruno, did not even know of him. It was not until nearly two centuries later that Schelling, the principal theoretician of natural philosophy, discovered him and made him known.

Baruch (Benedict de) Spinoza was a convinced rationalist, adopting a method of reasoning that was rigorously deductive, persuaded that he could thus remain in harmony with experience. He was rejected by the supporters of the Enlightenment for obscurantism but, in contrast, fascinated the German romanticists. For him God was unique and all-encompassing: He was neither matter alone nor pure spirit, but infinite substance, that is to say the essential reality underlying both matter and spirit and uniting them. In this sense, God was nature: *Deus sive substantia sive natura*, which is pantheism.

Spinoza's God was in no way transcendental. On the contrary:

"Whatsoever is, is in God, and without God nothing can be conceived" (Part I, XV of Spinoza's *Ethics*[10]).

"Besides God no substance can be granted or conceived" (I XIV).

"God is the indwelling, and not the transient, cause of all things" (I XVIII), which signified that God produced nothing outside himself. He was not really a creator God for "all things are conditioned to exist and operate in a particular manner by the necessity of the divine nature" (I XXIX), and therefore "Will cannot be called a free cause, but only a necessary cause" (I XXXII).

Of course, man also belonged to the divine substance and totality. It was therefore not surprising that Spinoza should see him in an uncompromising monist fashion. For him the two aspects of reality, matter and thought, body and soul, were not distinct, separate entities but rather the outside and inside of the same reality, even if the spirit was not reducible to matter. In consequence, "nothing can take place in that body without being perceived by the mind" (II XII), a very meaningful and almost prophetic affirmation when we remember that it took almost three centuries for the monistic view of body and soul to be accepted, not in pantheistic terms, but in the rational, empirical terms of modern neuroscience.

Science, for Spinoza, contributed to increasing man's power over things, hence the name *Ethics* given to his major work. He devoted many pages to man's bondage to desire and his liberation by reason. However, he left the scientist a proposition that Goethe, fascinated by pantheism, considered one of the most profound in the whole of literature: "The more we understand particular things, the more we understand God," for God and nature were one (V XXIV). In the end, was Spinoza a materialist or "intoxicated with God," *der Gottbetrunkene Mensch* as Novalis called him? In any case, he offered to the German romantics a natural mysticism that adapted itself very well to the constraints of reason, a lesson that was not lost on Goethe and Schelling.

The Lutheran Leibniz knew Spinoza personally. He respected him but hardly resembled him, except in his imperturbable belief in the power of reason—hence, the accusation of dogmatism made by Kant. He remained stony, if not arrogant, if anyone challenged him about common experience because appearances were just illusions related to human weakness.[11] Like Spinoza, he adopted a global philosophical view that emphasized the coherence among the divine, the human, and nature. In his work, he developed numerous concepts that were taken up and more or less modified by Herder in the second half of the eighteenth century and later by the German romantic and natural philosophical movements, which is why they are of interest to us here.

Everything depended on his metaphysical concept of the *monad*, Leibniz's major contribution, which he described with a conciseness and economy of effort that were as impressive as dogmatic:

The monad ... is nothing but a simple substance, which enters into compounds. By "simple" is meant "without parts."
These monads are the real atoms of nature and, in a word, the elements of things. (#1 and 3 of *The Monadology*[12])

Monads could only be created or destroyed by God; they could not act on the outside world or be influenced from outside:

Monads have no windows, through which anything could come in or go out. (#7)

Each monad was different and changed continuously according to an "internal principle." The activity of this internal principle expressed the monad's attempt to transform itself toward increasing perfection and was called *appetition* (#15).

Thus, the monad appeared as a force (in terms of Leibniz's dynamics) and as a hub of energy uniting matter and spirit. These successive transformations affected only a part of the monad; otherwise it kept its individual identity, thus preserving its invariance. This was the principle of continuity, which was important to Leibniz. This very principle allowed him to define the basis of transformism, which was later so precious for Goethe in his attempts to understand the history of the earth and its geological evolution as a succession of natural formations.

According to Leibniz's vision, all living beings, human or animal, represented an ensemble or community of monads, each corresponding to the above definition. So he had to resolve the important problem of the relationships of monads among themselves in each of these communities. Because monads depended exclusively on God, it was only through God that there could be communication and interdependence between monads (#51). The harmony of the whole was preestablished in God (#51–#55) in the best interests of the divine city of God's goodness (#86). Leibniz compared God to an expert watchmaker who had once and for all adjusted his monadic clocks so that, although independent of each other, they always chimed together, thus realizing a "pre-established harmony" (#78), a precocious psychophysical parallel that Gustav Fechner[13] developed later, in the epoch of romantic culture. Each individual monad, a hub of energy, was a "soul" (#19), but the ensemble of monads of a given living being was linked primarily to one of them, which became the dominant monad, and thus the soul of the man or animal. This prefigured the concept of the organism, so close to the hearts of romantic philosophers.

In animals, the soul, or dominant monad, was also provided with memory, which allowed them to recall previous perceptions: The sight of a stick, for example, would make them flee because it reminded them of the pain a stick had caused in the past (#26). As to humans, our "rational soul or mind" raised us "to the knowledge of ourselves and of God" (#29). Our consciousness, more complex than that of animals, could, however, be "stunned ... when there is a great multitude of little perceptions, in which there is nothing distinct" (#21). "And as, on waking from stupor, we are conscious of our perceptions, we must have had perceptions immediately before we awoke, although we were not at all conscious of them; for one perception can in a natural way come only from another perception" (#23). Here was a rousing entry of the subconscious into a story of which this was only the beginning.

The fact that the passage of information from one monad to another could only happen through the intervention of God led Leibniz to the conclusion that, "consequently everybody feels the effect of all that takes place in the universe. ... But a soul can read in itself only that which is there represented distinctly; it cannot all at once unroll everything that is enfolded in it, for its complexity is infinite" (#61). So when one monad increased, the others were obliged to decrease, giving the illusion that they were bound by a causal relationship, whereas everything could be explained by their harmonious coexistence.[14] Thus, one must support the notion of each monad containing the representation of all others.

Further, Leibniz was a confirmed vitalist when he affirmed that men's minds were "images of the deity ... capable of knowing the system of the universe, and to some extent of imitating it ... each mind being like a small divinity in its own sphere" (#83).

Finally, nature took on a cosmic and creative sense to such an extent that one could speak here, as in Spinoza, of a *Deus sive natura,* for "things lead to grace by the very ways of nature" (#88), and "God as Architect satisfies in all respects God as Lawgiver, and thus sins must bear their penalty with them, through the order of nature" (#89).

By introducing infinity in the finite and the unconscious in the conscious, by affirming the creativity of a nature inhabited by the divine and the concept of preestablished harmony, making of each monad a whole world like a mirror of God and of the whole universe, Leibniz weighed heavily in the heritage left by philosophers to the German romantics.

Natural Philosophy

It is difficult to imagine natural philosophy and romanticism separately because their psychological and ideological mechanisms were so intertwined. But if the passion of their union was one of the prime features of German culture, they did not escape serious frictions. For instance, Goethe had shaken off the main extravagances of romanticism when he published his scientific works, which were steeped in natural philosophy. The latter, moreover, underwent innumerable renaissances and avatars long after the extinction of romantic ardor.

Natural philosophy stood in contrast to Newton's quantitative, mechanistic vision, and the rigor of Kant's judgment, by its global, intuitive, and qualitative view of nature driven by the spirit. Schelling was its founding philosopher, and around him flocked numerous personages, often high in color: ecclesiastics and poets, physicians and physicists, all inspired by natural philosophy to varying degrees, but sometimes outrageously so. But he may never have become the philosopher we know him as today had there not been, thirty years before the publication of his works, the period of intense literary activity that was called *Sturm und Drang*, a sort of prologue to the era of natural philosophy and romantic culture.

This romantic literary movement, marked notably by Johann Gottfried Herder[15] and Goethe, expressed the unease of a generation in revolt against the preceding one, particularly against the inherent values of the Enlightenment, preaching instead a return to nature, the Middle Ages, and old popular values. However, it was far from being a generalized tidal wave and even less so a revolution. It ultimately derived its name from a play written in 1776 by Friedrich Maximilian Klinger,[16] a poet of a rather volatile nature. But for the historian, *Sturm und Drang* was born in 1770 (see below), lasted hardly more than ten years, and did not prevent the players of the Enlightenment pursuing their activities. In 1779 Lessing published *Nathan the Wise*, a eulogy of tolerance worthy of the Enlightenment, and in 1784 Kant wrote his essay on *What Is Enlightenment?*[17] It was nevertheless a key moment in literature because, for the first time, a German novel, Goethe's *Werther* (1774), took the whole of Europe by storm, even as far as Napoleon's bedside table.

Herder and Goethe in Strasbourg: Sturm und Drang

Strasbourg was then a French city, but its inhabitants were officially German subjects of the king of France. Its citizens endeavored to live in the French style, and that is why Goethe's father chose this university for his son. One day in September 1770, climbing the stairs at his hotel, Goethe bumped into a young clergyman who had turned up the hem of his coat by putting the corner in his pocket and wore his hair in round curls: He was an unusual young man but elegant and likable. Goethe recognized him as the young but famous Herder.[18] The year that this encounter took place has remained of symbolic importance because for historians it marks the real beginning of the short but intense *Sturm und Drang* movement.

Herder was in Strasbourg for treatment for his eyes. He was a theologian, educated in Königsberg, but he had rejected the rationalism of his teacher, Kant, in order to turn all his attention and enthusiasm to Johann Georg Hamann,[19] the "Magus of the North," for whom reason was just good enough to discover the disaster of birth, while it should be the servant of faith and genius, which could not be explained and without which nothing could be explained.[20] Equally for Herder, the spirit of the Enlightenment should be rejected: Life was merely a long Shakespearian drama, and there could no longer be any question of assimilating nature to reason.

He possessed a mind of great breadth and depth, with an encyclopedic knowledge of the history of peoples and their beliefs. He was convinced that all cultures worthy of the name emerged from ancestral popular traditions, the soul of the people (*Volksgeist*). It was the nation that permitted the blossoming of genius because true creative forces were collective, and poetry was their most ancient and authentic expression. He had an almost "biological" vision of the growth of societies, and, whereas Lessing spoke of the *education* of men, Herder allotted to the philosopher the mission of *giving life*, of being nature's physician, as it was constantly created, ordered, and destroyed."[21] In 1773 he published with Goethe and others "*Of German Character and* Art," which dealt in particular with the beauty of popular song.[22] Thanks to Goethe's intervention, he was appointed superintendent of the reformed churches in Weimar in 1776.

Born in 1749 in Frankfurt am Main, a year before the death of Johann Sebastian Bach, Johann Wolfgang Goethe died in Weimar in 1832, just as Hector Berlioz was finishing his *Symphonie fantastique*. He was both actor in and privileged witness of a period that counted among the most troubled, but also the richest, of European culture. Exceptional as a prose writer, but especially as a poet, with a universal curiosity and a broad and often visionary culture, at the end of his life he nevertheless attached more importance to his scientific than his literary work. What still surprises us today is that his scientific knowledge was based essentially on the broad concepts of natural philosophy. At the time he published the results of his research he had already abandoned the romantic literary movement and had clearly distanced himself from Schelling. An excellent speaker, he tended to be rather conceited: In a letter to his sister, he once compared girls to monads, who did not know that there were "little animals which could dance a minuet upon the point of a needle."

At the age of 21, he arrived in Strasbourg from Leipzig where he had been living a rather dissipated university student life, alternating writing poetry, philandering, and lectures on civil law. By then he had become a little more thoughtful, always sure of his own values, but desirous of formulating a European identity. At university he followed a mixed bag of courses in science, anatomy, surgery, history, and chemistry. His plan to take a doctorate in law fell through, and he had to be content with just a basic degree. It is not surprising that his proposed dissertation, *The Power of the Legislator to Determine Religion and Culture*, in which he opposed prevailing orthodoxy, was refused by the Faculty of Law. One of his minor theses was discussed but judged simplistic by the faculty. Titled *Natural Law Is What Nature Teaches All Animals*, it clearly anticipated his conception of romantic science, in which nature gave lessons of life and that science must interpret instead of putting itself in the service of technology.[23]

Long discussions with Herder, whose influence he admitted and who literally hypnotized him by his eloquence about the history of nations, poetry, genius, and the soul of the people, greatly helped him mature. His walks through the countryside of the Vosges, during which he collected popular songs from elderly Alsatian women, his fiery romance with Frederike, and the daughter of the pastor of Sesenheim all induced him to live intensively with nature. "Nature is a melody in which is hidden profound harmony," he wrote. From pietist, in agreement with their highly emotional approach to religion, he became definitely pantheist, and he developed an unreserved admiration for the great Spinoza.

He came to Strasbourg to be close to French culture, but he rebelled and felt the awakening of the dormant German instinct in himself. The first time he visited Strasbourg cathedral, he said: "I had my head full of general knowledge of good taste. Through hearsay I respected harmony of volume and purity of form, but was the declared enemy of arbitrary and bizarre gothic ornamentation." He changed his opinion a few months later and proclaimed all his admiration for the architect Erwin von Steinbach. A so-called "expert" who denigrated his work by calling this cathedral *gothic* should "thank God to be able to proclaim aloud that this is German architecture, our architecture."[24]

Strasbourg cathedral had long been deeply symbolic for French and German intellectuals. In his 1836 essay, *The Romantic School*, Heine, faithful to the Enlightenment, dreamed: "Oh, would that I could stand on the cathedral at Strasbourg, a tricolor in my hand that would be seen as far as Frankfurt. I think that if I waved the flag of my dear homeland and proclaimed the right words of exorcism, the old witches would fly away on their broomsticks ... and all the apparitions would cease."[25] This passage was of course deleted by the German censors of the time.

Back with his family in Frankfurt, Goethe wrote his *Werther* (1774), a novel full of melancholy and despair, and *Götz von Berlichingen* (1773), in the image of the popular hero struggling against social injustice. By this time, famous in his own right, he was appointed Privy Councilor to Karl August, Duke of Weimar. It was in this Athens of the north, this highly cultural city, where Christoph Martin Wieland, novelist and promoter of a national poetry, was already and where later Herder and Schiller would arrive, that Goethe spent the essential part of his existence. There, having definitively rejected all romantic excesses, he developed a desire for more classical cultural values. He pursued his work on *Faust* and relentlessly wrote novels, poems, and plays. He laid the foundations for his scientific research true to a metaphysical vision of nature: geology, osteology, botany, and, above all, his theory of colors, of which we shall explore the essential themes and contrast them with those of Helmholtz, the physiologist.

Nature-Object or Nature-Subject?

One of the essential keys to understanding Goethe and the development of German romanticism and natural philosophy is to be found in a certain conception of nature: nature created or nature creator? Nature-object or nature-subject? Created nature was that of Galileo or Newton, and later that of Helmholtz, an object of knowledge, subject to predetermined mechanisms that could be formulated quantitatively by means of physicomathematical models. The other was sovereign divine nature that conceived and engendered all things: Spinoza's *Deus sive natura* to which man belonged or Leibniz's constellation of monads, where any modification of activity of any of them had repercussions on the activity of the others. Man could thus influence nature, and vice versa, in the context of a harmony preestablished by the divine monad that encompassed everything. This nature-subject, which ensured the fertility of the earth or triggered catastrophes and epidemics, and which ordered life and death, had been feared and respected since the origins of man, who, to ensure its benevolence, organized phallic or dionysiac rites.

A very large part of the German nation, perhaps even the majority of its vital force, at the end of the eighteenth and beginning of the nineteenth centuries, believed in the creative force of nature and its close relationship to man. There were several exceptional minds who incarnated and expressed this belief and who made the Germans more aware of their particular spirit. Had not Herder, the theoretician of the spirit of the people, exhorted his compatriots to be nature's physicians? As to

Goethe, he sang the praises of the "profound harmony" of nature, as both a poet and a scientist. His poems are like an analogy of his scientific investigations, and that is their special interest. The logic underlying his poetic expression makes it fitting for us to pause for a few moments ... time enough for a poem. Goethe wrote this one autumn evening in 1780 on the wall of a mountain cabin in Thuringia:

Wandrers Nachtlied:
Über allen Gipfeln
Ist Ruh,
In allen Wipfeln
Spürest du
Kaum einen Hauch;
Die Vögelein schweigen im Walde.
Warte nur, balde
Ruhest du auch.
[Wanderer's nightsong:
Over all the summits
Is peace.
In all the tree tops
You hardly feel
The slightest breath;
The birds in the forest are silent.
Just wait, soon
You will also rest.]

The first thing to notice is the remarkable musicality of the German poetry that cannot be reproduced in a translation, however good. Young Goethe generally wrote his poems about nature by alternating the relationship between subject and object.[26] In the first six lines, nature plays the role of subject, and the interlocutor of Goethe is the object. However, the situation is reversed in the last two lines, where the interlocutor becomes, in turn, the subject. This moment of reversal of the relationship, this demarcation—an essential concept in the perception of colors according to Goethe—is abrupt between the two parts of the poem, where the poet, by using the familiar form "*du*" for "you," plays the role of intermediary guaranteeing harmony and the unity of man and nature.

But there was another domain that revealed how deeply rooted in creative nature the German mind was: that of law, and the astonishing conversion of natural law derived from the Enlightenment into romantically inspired law. Samuel von Pufendorf and Christian Thomasius had espoused the French preoccupation since the end of the seventeenth century for constructing a natural, rational law that was, by its very universality, above any state. However, Karl von Savigny,[27] professor at the University of Berlin and future Grand Chancellor of Prussia, rejected this as a pure product of human reason, and therefore arbitrary. He stated in 1816 that universal man did not exist in reality and that law stemmed from common legislation, expressed the spirit of the people and evolved at the same time as it.

Jena: Birth of Natural Philosophy

The *Sturm und Drang* period was merely a preromantic episode, tumultuous but soon exhausted. What we usually call romanticism only saw the light of day later, around 1795, when the two friends Schiller and Goethe had become frankly classical. The former wrote *Wallenstein* (1800), and especially his *Letters upon the Aesthetic Education of Man* (1794), in which beauty became the reflection of liberty in the world of sensations. The second finished *Wilhelm Meister's Apprenticeship* (1795), the portrait of a man integrating himself in a harmonious society. This last work, above all, provoked the anger of Novalis,[28] the first of the romantics, who found it "hostile to poetry." Novalis, whose real name was Friedrich von Hardenberg, was the emotional image of a poet "hoping for an imminent joyous end," which indeed he achieved before the age of thirty. He had seen in his dreams the "blue flower," the folkloric image that Herder attributed to Hindu mythology, and that signified for Novalis that in dreams where it rained wine and snowed roses, opposites were not contradictory.[29] The more something was poetic, he said, the more it was real. One needed the simplicity of a child to study nature. With this Novalis established a true charter for natural philosophy as the poetry of nature.

It was at Jena, a few kilometers from Weimar, that the originators of romanticism met. In addition to Wilhelm Schlegel and Schelling, both professors at the local university, there were Novalis, Johann Wilhelm Ritter, Friedrich Schlegel, Ludwig Tieck, and others. The pope of literature was not far away, and it is said that they danced around Goethe like children around a Christmas tree.[30]

In this profoundly romantic context, and probably under the influence of Novalis, Schelling undertook the task of unifying knowledge of nature and making it coherent, which made him in the eyes of posterity the founder of natural philosophy. A student of Fichte and appointed at the university thanks to his and Goethe's intervention, he soon detached himself from his master's Kantian line of philosophy and fell into conflict with him. His great project was to reconcile the finite and the infinite, an old problem already posed by Bruno. At first, like Fichte, he claimed that "self" was the only authentic absolute, and he expected that in the new philosophy it play the role that God played in the dogmatic philosophies of Descartes, and especially Spinoza, that of the principle of absolute knowledge.[31] It followed that anyone confronted with nature could describe it according to his own model and derive those principles necessary to explain it from his own experience. The nature of the human mind was the final explanatory key for nature itself. We may recall this fine romantic statement by Schelling in the introduction to his *Ideen*: "Nature is visible Spirit; Spirit is invisible Nature,"[32] which resolved the irritating realist question about nature outside us in the light of nature within us.[33] This explained the philosopher's intuition for a joint effort from productive forces and vitality, permitting a progressive dynamic synthesis of nature and man thanks to the universal sprit of nature little by little shaping the raw material to its own likeness.[34]

Schelling set himself the fundamental task of establishing a channel of ascensional intelligibility from the origins of creation to man, and even beyond man, in which the doctrines of survival and palingenesis, so dear to the Gnostics, found all their significance.[35] Thus, the evolution of nature was marked by ever-increasing value, and creation would never be completed. This explained the importance that he attributed to a "return to origins" necessary for unraveling the various forms revealed in nature, and which found an echo in the obstinacy of Goethe's search for the *Urpflanze*, the primeval plant of which the form was the archetype of all subsequent plants.

Another concept that was dear to Schelling was polarity. It was the basis for the creative dynamism of nature, which depended on the conflict of two antagonistic forces that tended to equilibrate without ever succeeding and of which the magnet was the best example achieved by nature.

Schelling the idealist thus constructed the philosophical supports of romantic culture and speculative science, and numerous poets, physicists, and physicians eagerly joined the debate. To consider them all would be tiresome, but we must cite a certain number of their theories and aphorisms that allow us to better understand the consequences of this line of thought. For example, experimental demonstrations had become pointless. There was an almost Leibniz-like preestablished harmony between thought and nature, as both were identical. This allowed Novalis to say that if theory had to await confirmation by experiment, it would never reach a satisfactory conclusion.[36] Experimental physics could therefore never be more than a juxtaposition of dispersed analytical works without any possible link to the obvious unity of nature and man's place in the universe. All that counted was "speculative physics," which allowed one to consider and justify nature as preexisting man and to give it some sense, as vision was the sense of the eye. For Schelling, "we do not know nature for nature is *a priori*." The result was that natural science became a deductive and demonstrative philosophy.[37]

Thus, natural philosophy lay at a higher cultural level, at a different level of intelligibility from that of empirical science, that overhung an unfathomable abyss.[38] So it was not surprising that the rejection of experimentation was also accompanied by the exclusion of mathematics. Not only Goethe, but even Schopenhauer, for whom where calculation began comprehension ceased, shared Schelling's opinion denying mathematics any say in the matter of higher physics. On the other hand, arithmosophy had its place because most natural philosophers were searching everywhere for the same numerical structures in crystals, constellations, the blood circulation, and the periods of human life.[39]

Finally, one must recall the importance in natural philosophy of the concept of the organism, which Leibniz had developed in his *Monadology*, emphasizing that the diversity of a given animal's monads in no way prevented one of them from becoming dominant and so becoming the animal's soul. The organism was a perfect model of "sameness in difference." According to Schelling, there had to be a principle reproduced in all parts of the whole, an organic unity. The soul of the world (*Weltseele*) was the principle uniting nature in a vast organism.

One cannot help being struck by the complexity of the natural philosophers' vision with its impression of wanting to embrace nature with such passion that they could only express themselves in a speculative, inspirational language that was in turn poetic, religious, philosophical, or physical. What was more, the concepts they borrowed from physics, such as magnetism and galvanism, only found their full expression in the world of symbols. As for science, it no longer needed experiments, except as illustrations to convince skeptics, for theory was the priority. The only observations that counted were those of some significance in the context of their theoretical vision of the universe. Goethe, whose main research was carried out before the writings of Schelling and the romantics, largely anticipated them, but with more subtlety.

The Esoteric Meanderings of Theosophy

If you enter the cathedral of Siena through the bronze doors of the nave, you are immediately welcomed by the portrait of Hermes Trismegistus inlaid in black in the white marble of the floor (figure 2.1). It is attributed to Giovanni di Stefano, son of Sassette, and dated to 1488. Hermes was the Greek avatar of the Egyptian god Thot, god of the Nile delta usually represented as an ibis or baboon. A mythical sage and contemporary of Moses, he invented language and writing, science and art. He was scribe to Osiris and entrusted by the latter with the calculation of time and the mastery of magic. Hermes, thrice great messenger and spokesman of the gods, whose words Plato said he imagined in his *Cratylus*, became in the Hellenistic period the supposed author of books on alchemy, astrology, and mystic philosophy. In hermetism, science and religion were not separate. Knowledge was always revealed and was the result of asceticism and piety rather than rational thought, which ran counter to the views of Aristotle.

The Siena Hermes surprises us by the eloquence of his message. The white of his clothing symbolizes illumination and innocence but also the element mercury, placed low in the material microcosm. Mercury is also the planet and the Latin name for Hermes. The planet Mercury is placed high in the macrocosm of the celestial bodies. Their reciprocal positions permitted them to regulate the activities of the spirit descended on earth. Astrology was firmly linked to alchemy.[40] This astonishing monument in the heart of catholic Tuscany illustrates both the intellectual imagination and the philosophical hesitation of the Renaissance. Indeed, a return to antique thought, especially to Plato, marked this period. Aristotle still had numerous disciples among those who recommended a closed, eternally determined, hierarchical system for the cosmos, but Plato was preferred because he associated physical man with more personal responsibilities in the matter of individual ethics, and he preferred a search for truth rather than the acquisition of certainties. This usually resulted in a struggle between these two philosophies, but also in attempts at reconciliation between them, as in the *Oration on the Dignity of Man* by Giovanni Pico della Mirandola[41] or the fresco of *The School of Athens*, where Raphael represents Plato holding the *Timaeus* in his hands, the most Aristotelian of his works, and Aristotle the *Nicomachean Ethics*, his most Platonist work.

In fact, this abundance of philosophical imagination inspired by antiquity also indicated a certain confusion as a result of the changes to mankind after the discovery of the New World and the use of mass

Figure 2.1
Portrait of Hermes Trismegistus in the floor of the cathedral of Siena, attributed to
Giovanni di Stefano, 1488.
(Scala/Art Resource, NY)

communication techniques, such as printing. The desire for progress in knowledge was universal, but Copernicus and Galileo had difficulty finding support for their heliocentric concept of the planets because science was at its very beginnings, and its usefulness was far from accepted by all. This explained the interest of so many intellectuals of the time, after Marsilio Ficino[42] and Pico della Mirandola, for esoteric and hermetic visions of knowledge and interactions among God, the stars, and man.

The advances in science that began with Galileo did not immediately put a stop to this sort of vision. Johannes Kepler, who calculated the trajectory of Mars and discovered its elliptical orbit round the sun, was nostalgic about the unity of physics and metaphysics and attributed a soul to the sun, sensitive to musical harmonies. Even Newton published *Praxis* in about 1693, a text on alchemy in the purest esoteric tradition.[43] Having said that, the confusion caused at the Renaissance by the loss of Aristotelian values and the absence of a scientific culture worthy of the name doubtless favored credulity in esoterism, astrology, and magic, beliefs inherited from the depths of antiquity. For many intellectuals, the "divine analogy," the divine law that governed all beings, was preferable to the experimental method, remote from any religious context.[44]

We saw earlier how much natural philosophy owed to philosophers such as Bruno, Spinoza, and Leibniz. But certain trends in esoterism, better known after the Renaissance as theosophy, equally played an important role in its genesis. Theosophical thought was steeped in the mysteries of the nature of a God that was found everywhere. It was symbolic, analogic, and creative; used alchemistic and arithmosophic concepts; and everywhere sought for polarities of opposites. It began where rational philosophy ended, and it ended where theology began. The notion of theosophy, where to understand things one started with God, embodied pansophy, especially since Paracelsus[45]: To accede to knowledge of the divine, one began with deciphering the messages of nature and the concrete world. The natural philosophers, just like the alchemists, were true pansophists.[46] This concept of theosophy found a particularly fertile ground in Germany, where numerous were the philosophers who adopted the approach. The physician Paracelsus, for example, was a truculent and colorful character, whose real name was Theophrast Bombast von Hohenheim. He was from an old Swabian family and was a particularly representative figure of theosophy. We owe him numerous medical discoveries, such as the link between goiter and cretinism, silicosis and work in the mine, but above all his practice of iatrochemistry and his guidance of alchemy toward therapeutics. A prolific and humanist writer, he took his inspiration from the work of Ficino and Pico della Mirandola and popularized Plato and the Kabbalah.

Protestant Germany of the seventeenth and eighteenth centuries had certainly not forgotten Paracelsus, but they also remembered an ancient theologian from Thuringia, Meister Eckhart,[47] who had taught that, far from being immutable, God was in constant evolution, and he needed nature and its creatures to exist.

Two major theosophers adopted a large part of this heritage and marked Lutheran thought. First, Jakob Böhme[48] developed the theme of original freedom, a sort of dynamic obscurity from which stemmed spirit and nature, caught up in the play of the inseparable forces of positive and negative. Then Friedrich Christoph Oetinger[49] apparently had a great influence on the romantics. He favored the use of experience and perception rather than mathematics, and thus he sided with a certain degree of empiricism, which was part of a global vision that excluded any form of dualism. God and the world were interlocked, and, as Faivre said so elegantly,[50] Oetinger saw in nature a "great academy," the humblest thing bearing witness to God's "invisibilities."

There is little doubt that theosophical thought found an echo, and sometimes even more, among the natural philosophers. Some of them, such as Novalis and Franz von Baader, were true theosophers. Schelling, in contrast, certainly did not consider himself a theosopher, but he had read and appreciated Baader, who was his university colleague.[51]

Natural Philosophy and Medicine

Natural philosophy had a considerable influence on different aspects of German science and art, including medical science. In 1840, despite the development of experimental science, this influence was still very tangible. Even the aging Alexander von Humboldt, who had supported Schelling's early career, expressed some very hard thoughts on the subject of natural philosophy, saying that the abstract claim of completely false facts and opinions signified alienation and anguish.[52] The relationship between medicine at the beginning of the nineteenth century and natural philosophy merits more than just a passing remark. Indeed one must remember that at this time medicine was far from establishing itself; it was still struggling to find a coherent and stable scientific doctrine. Medicine as taught in the major universities was mostly bogged down in authoritarian immobility. The hospitals, the only places where any sort of practical experience could be had, were dangerous sources of infection. It is significant that Thomas Jefferson, the third president of the United States, stated in 1806 that "Harvey's discovery of the circulation of the blood was a beautiful addition to our understanding of animal economy, but on a review of the practice of medicine before and since that epoch, I do not see any great amelioration which has been derived from that discovery."[53] Furthermore the historian Charles Lichtenthaeler[54] did not hesitate, in one of the chapters of his book, to deal with the chaos in medicine in Paris at the beginning of the nineteenth century. It was only in 1816 that René Laënnec,[55] one of the fathers of modern clinical medicine, invented the stethoscope, which opened wide the objective study of cardiopulmonary diseases, while François Magendie[56] was imposing his experimental approach, claiming that "medicine is the physiology of the sick," thus preparing the way for Claude Bernard.[57]

As in other countries, medicine in Germany at the beginning of the nineteenth century was striving for scientific coherence, but what differentiated it radically was its allegiance to Schelling's natural philosophy, which made some sense of its efforts. Of course, before Schelling there was Kant, whose complex message had made a profound impression on scientists in general and physicians in particular. Kant's influence was central to the awakening of German physiology, but it was not really felt until after 1810 when romanticism and natural philosophy began to wane.

Schelling and Medicine

It is not surprising that Schelling's ideas caused a stir in the philosophical backwoods because they expressed, as abruptly as spectacularly, the passage from the empiricism of the Enlightenment to the often heteroclite speculations of the romantics. Nevertheless, Schelling brought to the physicians intellectual tools that allowed them to conceive a global view of man struggling with disease and death. This helped them, in the face of their lack of more elaborate scientific knowledge, to better understand and care for their patients. The message of the founder of natural philosophy was not far removed from Goethe's ideas: The world was unified because thought and nature were identical. Philosophers and physicians must strive to demonstrate the reality of this unity: unity of body and mind, conscious and unconscious, man and nature, organism and environment. Man, like nature, was a seat of tensions between opposite forces and polarities: attraction and repulsion in mechanics, acids and alkalis in chemistry, negative and positive in electricity and magnetism. In the end, man and nature were the seat of continuous progress, which ensured their development from their respective archetypes toward a final flourishing unity.

It is therefore also not surprising in this context that the nosological concepts of John Brown met with broad favor by the German medical body. This Scottish physician had been impressed by the research of Albrecht von Haller, the renowned Swiss poet, surgeon, and botanist,who taught in Göttingen. The latter had made a distinction between "irritability," the reaction of living tissues to outside stimulation by a contraction, and "sensitivity," the property of nerve fibers to evoke pleasure or the contrary. Haller interpreted Brown's observations very personally, claiming that life was not a natural process and that it only existed as a function of an external stimulus producing an active and lasting reaction. In such a definition of life, one no longer attributed the idea of a vital force to an active principle (a metaphysical cause), but that of a defensive force.[58] Brown's insight led him to try to define these forces, which he did from his observations on the sick. Sickness was only different from health, he said, because of an imbalance in the forces of defense: The sick man was either in a *sthenic* state marked by an exaggerated reaction of the organism or an *asthenic* state marked by a weakened reaction. The treatment was to restore the balance by calming or stimulating reactivity. Various medications were indicated: Opium notably became the romantic drug *par excellence* for treating languid diseases such as phthisis, and it was consumed in the form of laudanum by intellectuals.[59]

John Brown's medical theories were introduced to Germany around 1795 and received the favor of Schelling, who saw therein the result of a theoretical effort far removed from empiricism. This was also true of his friend Johann Andreas Röschlaub,[60] an important clinician of the romantic period who contributed to ensuring their publication. Schelling did not hide his supposed medical expertise. One day, with the agreement of Röschlaub, he administered opium to his daughter-in-law, Augusta, who was suffering from serious dysentery. She died, and it caused a major scandal.

In the same romantic vein that situated the healthy and the sick in the same global perspective, one should also mention the homeopathic doctrine of the Saxon physician Samuel Hahnemann.[61] This doctrine, originating in 1786, was not really an offspring of natural philosophy because it was more empirical than theoretical, but it nonetheless pursued certain

common objectives and dealt, like natural philosophy, with the relationship of macro- and microcosms. In the tradition of Hippocrates and
Paracelsus, Hahnemann adopted the principle of similitude: That which
made an individual sick could also cure him.[62]

The medicine of Brown and Hahnemann both had a positive influence
on therapeutics, in that the sick were better protected against the excesses
of traditional medicine, with its blood-letting, emetics, drastic purges,
and even more dangerous drugs, although Hahnemann had far less
success during the romantic period than Brown. Brownism, despite
its lack of a scientific basis, was thus not the catastrophe that many
described subsequently, when medicine had developed thanks to the
progress of experimental physiology. In her well-documented work,
Lohff[63] protested against decrying the medicine of the romantic period,
which was said to have been under the spell of the dogmatism of natural
philosophy until, suddenly after several decades, scientific medicine was
reborn like a phoenix from its ashes. On the contrary, the combined
intellectual activity of philosophers and physicians of the period actively
prepared the advent of methods and problems that were suitable
for experimentation.

But there is another aspect of the medicine of the romantic period
that we should emphasize. This was the rather beneficial influence of
natural philosophy on the relationship of the doctor with his patient. The
patient was above all a person, a whole person who the disease afflicted
as much in his psychic equilibrium as in his physical well-being. His
symptoms had to be interpreted in the context of their relations with the
familial and natural environments. Even more so because these relations
were reciprocal and, in the end, man was in relation with the whole universe: Since Hippocrates, it was considered that the equilibrium of the
microcosm followed that of the macrocosm. Was not the physician a
metaphysician who considered the universe as a vast organism? For the
physiologist Karl Friedrich Burdach,[64] the practice of medicine was like
the priesthood: "The world ... is the absolute organism in which the
primeval spirit (*Urgeist*) is expressed, the infinite and eternal revelation
of God."[65]

The romantics imposed a human face on medicine; thanks to them, it
became anthropology applied to pathology.[66] The influence of Novalis
was very important in this respect. Long-lasting illnesses had to be lived
like the call of destiny, nature's call to order. They were years of apprenticeship of the art of living and a school of the inner being. As to the
doctor, he sometimes resembled a magician. Disease was a musical

problem, and a cure was a musical solution. The greater the musical talent of the doctor, the quicker and more complete the solution.[67] A characteristic was the relation of the doctor to death, which he did not perceive as an irremediable failure of his treatment. Indeed, he helped the patient accept death as a natural phenomenon that no one could avoid and that one should try to accept not as the end, but as an achievement or the beginning of a new life. Death was not without meaning and could even have a frankly positive significance.

Natural Philosophy and Physiology

One might think that such medicine, more preoccupied with wanting to transmit its philosophical concept of the experience of disease and death than with research for effective therapeutic methods, might completely neglect physiology and experimentation. But it was not so. The major themes of physiology were dictated by contemporary preoccupations.[68] On the one hand were the medicine of Brown and questions concerning excitability and the balance between stheny and astheny. On the other hand was the vital force that differentiated the organic world from the inorganic: material or immaterial? How were the vital forces of particular organs regulated? Vegetative, reproductive, neuromuscular forces and others still?

But what could physiology really do in the context of a romantic culture and a theoretical philosophy of nature? Certainly not research into a continuous chain of successive causes going back into the indefinite past and incapable of solving a given problem. On the contrary, physiology, said the philosophers of the time, had to tend toward becoming "zoonomia,"[69] that is to say a doctrine of life in which comprehension of animal life was seen as a function of the laws of philosophy.[70] Of course the physiologist was invited to observe facts and to experiment, but on the condition that he respect the unity of body and mind, that he consider the organism as the blood of the universe, and that he reconcile empiricism with the theory and context of the dogmas of natural philosophy. Indeed, for Schelling the aim of science was not to describe nature's products but to reconstruct *a posteriori* the processes of their elaboration by a theoretical approach in which nature was considered as incessantly active and permanently productive. More generally, the objective of "speculative physics" was to unite the organic and inorganic in a single common expression.[71] Nothing was superior to reason, Schelling reminded. It was not simply a tool that we used; rather, it was reason that possessed us. It was the equivalent of the spirit of nature and therefore could not deceive us, which justified the idea that the empirical approach of the physiologist had as its only aim the illustration and confirmation of theory.

At first sight, the balance of influence of natural philosophy on the development of physiology was frankly negative. Although the latter certainly occupied a dominant position in German culture, it did not create conditions propitious for the growth of new empirical knowledge. The result, according to certain German scientists and historians, was a dramatic reduction in scientific progress after the Enlightenment. This unequivocal judgment was prevalent from 1840 onward and even seemed to worsen during the twentieth century.[72] However, in the last twenty years or so, there has been a certain revision of such intransigent views in Germany, and new, subtler opinions have emerged. For example, Lohff[73] did not believe that the romantic period was marked by a scientific vacuum. On the contrary, she represents the history of physiology during the last two centuries as a voluminous treatise of which the romantic scientists wrote the preface and introduction, after which the postromantic and present-day researchers assumed the task of writing and progressively enriching the later chapters. The pages written by the romantics were reflections on themes directly related to physiology, such as the unity of life in the diversity of its manifestations, the limits of physiology as an explanatory science, the relations between theory and empiricism, the relative character of observation in the experimental method, and the theory of knowledge. In its quest for scientific credibility, the physiology of the romantic period established a conceptual context for its research domain and asked a number of questions. To neglect such an introduction would leave the other chapters of the work without coherence. This part of the history of physiology was necessary for its overall understanding. Indeed, not taking notice of it led to physiology overestimating itself and disappointing false hopes placed in it.

The physiologist Johannes Müller experienced very closely certain excesses of natural philosophy, and all credit is due to him for having resolutely put his research on an empirical basis. His *Habilitation* speech in 1824 had indeed historic significance because he endeavored to reconcile the contributions of "good" natural philosophy with the necessary rigor of experimental physiology, as we shall see in chapter 3.[74] It is also not devoid of interest to read a few lines from a speech by Helmholtz in 1878, then age 56, at the University of Berlin. He was at the summit of his fame as an empirical scientist, and he had never hidden his aversion to the errors of natural philosophy and the idealist philosophers who had introduced metaphysics into science. He was alluding to Fichte, former Rector of Berlin University, but doubtless also to Schelling and Hegel. "Even the aberrations of this way of thinking, expressed in the obvious weaknesses of the romantic movement, have something attractive compared to cold, calculating egotism. One admired oneself in the fine sentiments in which one wallowed; one sought to create art that could give such sentiments; and one felt able to admire fantasy as creative art the more one distanced oneself from the laws of reason. There was much vanity in this, but vanity that always knew how to be enthusiastic for high ideals."[75]

Kant and Empirical Science

One should not think that during the whole period when natural philosophy dominated in Germany, Kant, the philosopher of the Enlightenment, had been relegated to oblivion. On the contrary, physicians and physiologists alike had understood his message about the origins and criteria of scientific knowledge, and those theories were the subject of numerous discussions among themselves. However, the time was not ripe for an essentially empirical approach to science because of the influence of Schelling's philosophy and his speculative, deductive approach to science. But later, after the romantic period, many scientists saw in Kant an inescapable philosopher. This was the case for Helmholtz.

One of Kant's great merits was to have restored the principle of causality that had been so roughly treated by the English empirical philosopher David Hume. For the latter, the mind was reduced to a whirlpool of different perceptions one after the other. Speaking of relationships, such as that between fire and heat, for example, he said, "the knowledge of this relation is not, in any instance, attained by reasonings *a priori*; but arises entirely from experience, when we find that any particular objects are constantly conjoined with each other." Further, "we are determined by custom alone to expect the one from the appearance of the other.... Custom, then, is the great guide of human life."[76] So would science henceforth be reduced to mere observation of the probability of temporal relationships?

At this stage, Kant made his entrance, "awoken from his dogmatic somnolence,"[77] as he put it himself, by the profound skepticism of Hume and the death knell that was sounding for all science worthy of the name because the latter believed he had discovered that what one usually took for reason was merely an intellectual illusion. In his *Critic of Pure Reason*,[78] Kant hoped to demonstrate that the principle of causality really existed. It was not to be sought in the interplay of external objects, but in the properties of the human mind which, before any experience and thus a priori possessed the transcendental principles needed to dispose of a certain level of understanding. To access understanding of outside objects, the mind used two complementary tools: intuition and intelligence.

First, intuition (*Anschauung*) only referred to the manner in which we were affected by objects. An object acted on the sense organs producing a representation that gave rise to an intuition, consisting of an image of the sensory perception in the form of a priori space and time. It is important to note that Kant's a priori is a philosophical concept, not a biological one. Why a priori space? Because "one cannot imagine that there is no space. ... Space is a necessary *a priori* representation that serves as a basis for all outside phenomena. ... So it is considered the condition for the possibility of phenomena, and not a determination that depends on them." As to time, "it is nothing else than the form of inner sense, that is of our intuition of ourself and our inner state ... I can say in general, from the principle of inner sense, that all phenomena, that is all objects of the senses, are within time and so are necessarily related to time."

Next, intelligence (*Verstand*) was the power to think about the object of sensory intuition. Thought lived on intuition thanks to rules or logical categories that Kant borrowed from Aristotle. They too were a priori because they were necessary, universal foundations for all experimental knowledge. Kant answered Hume, who had reduced the principle of causality to a psychological mechanism of habituation resulting from repetitive associations during a given experience, by saying that the principle of causality was a philosophical a priori for any experience, and that it was indispensable in order to experience anything. Anything that happened presupposed that it followed something else. Without such a precept, science could not exist.

Knowledge could only be achieved by the joint action of intuition and intelligence because intelligence could not be intuitive, nor could the senses think. Furthermore, "thoughts without content are empty, intuitions without concepts are blind." But if knowledge began with the senses and then progressed to intelligence, it nevertheless ended in reason. Reason unified a priori the rules of intelligence, which it steered toward a certain truth, of which intelligence had no concept. This truth tended to bring together in an absolute whole all aspects of intelligence that concerned a given object. Reason thus directed intelligence in its work and had a regulatory and heuristic role. But perhaps above all, the principal function of reason was moral because it was from a moral standpoint that we could give their true meaning to the ideas of the intelligent world.

It was thus not surprising that the problem of the laws of nature was posed very differently by Kant compared with Descartes. For the latter, the origin of these laws was the God of theology, whereas for Kant nature could only be seen in terms of categories of intelligence applied to sensory intuitions: Its laws were therefore those of our intelligence that allowed us to confer on certain generalizations of an empirical origin the character of necessity and universality. Consequently, the laws of nature were essentially mental constructs that were useful to deal with experience, and we ourselves introduced order and regularity in the phenomena that we called nature.

So from Descartes to Kant, theocentric rationalism evolved into anthropocentric rationalism. Empirical science found a philosopher who gave it a real status, and we shall see later the benefit that Helmholtz the physiologist obtained from the Kantian vision for his research on human perception. On the other hand, Kant never hesitated to state that science could not exist without mathematics, which gave an advantage to the physical sciences and left the physicians in deep distress, being so far from meeting Kant's criteria. The physician saw himself more and more as an artist rather than a scientist, which explains at least in part the success of natural philosophy among them.

3 Johannes Müller, "Man of Iron"

Gehalt ohne Methode führt zur Schwärmerei,
Methode ohne Gehalt zum leeren Klügeln,
Stoff ohne Form zum beschwerlichen Wissen,
Form ohne Stoff zum hohlen Wähnen.

[Content without method leads to fanaticism,
Method without content to empty subtility,
Matter without form to onerous knowledge,
Form without matter to hollow fancy.]
—Johann Wolfgang Goethe[1]

These verses were cited by Johannes Müller on the reverse of the cover of his *Zur vergleichenden Physiologie des Gesichtssinnes*.[2]

Born in Koblenz in 1801, the year that the Treaty of Luneville decided its attachment to France as capital of the department of Rhine and Moselle, Müller began his university studies in Bonn in 1819. But before registering, he spent several days wondering whether he should read law, as his friends advised, or theology, as his mother wanted. After mature reflection, shut in his room, he opted for medicine, which would allow him, as he said, to know what his hands were capable of and whom he would be serving.[3] The teaching he received was impregnated with natural philosophy. Speaking of physiology professor Philip Franz von Walther, whose works Müller incidentally later recommended to his own students, the historian Rothschuh[4] said that at that time not to express oneself in the language of natural philosophy was pale and retrogressive.

Müller attracted the attention of the university authorities by taking the first prize in a competition organized by the Faculty of Medicine on the occasion of the king's birthday to reward the best experimental research to determine whether a fetus breathed in its mother's uterus.

He showed great merit. Not only did he have to pay for his experimental cats and rabbits from his own meager resources, but he crowned his work with a real highlight: the dissection of a live pregnant ewe in which he demonstrated the arterial and venous blood of the fetus with different colored dyes. He showed that, despite the lack of respiratory movements, there was evidence of unequivocal respiratory-like activity. This was obviously explicable by gas exchange in the placenta between maternal and fetal circulations. He paraded through the town with his friends shouting, "Eureka!"

The following year he published his research on the movements of animals, notably insects and spiders, which led to the award of a precocious doctorate in medicine in 1822. This research, much of which was conducted with excellence, was in the field of comparative physiology. Curiously, he succumbed to the fashion of including in his text numerous references to themes dear to natural philosophy, such as defining life at the origin of movement as an organic column of which the poles were flexion and extension, references that he regretted later, according to du Bois-Reymond,[5] and that he expurgated from his thesis, after it had produced the desired effect of pleasing his jury. Had Müller given in to fashion, to a sort of intellectual terrorism in the environment, or did he really believe what he wrote? The truth is probably something of both.

The young doctor of medicine wanted to continue his training and had the chance to leave Bonn for the University of Berlin to join Professor Carl-Asmund Rudolphi,[6] the most renowned anatomist and zoologist in Germany. He was not able to practice experimental physiology because his mentor did not like to work on living animals, there being no anesthesia available at that time. On the contrary, he was plunged into the study of systematic zoology and especially comparative anatomy, particularly that of worms and other invertebrates.

Rudolphi was a rigorous scientist who mistrusted philosophical speculation. Indeed his influence was salutary for his student who, after his stay in Berlin, avoided confronting the problems of natural philosophy out of context, always trying to clearly separate them from his scientific research. What is more, his mentor soon became a true friend for young Müller. Rudolphi was a man, he said, whose personal qualities were no less than those as a scientist: "*integer vitae sclerisque purus*" (upright in life and free of sin). When he left to return to Bonn, Rudolphi gave him a microscope, a significant gift that was to weigh heavily in his scientific evolution.

A few days later in 1824, he presented his *Habilitation* thesis to his Faculty of Medicine, titled *The Need in Physiology for a Philosophical Reflection on Nature.*[7] Nothing remained in it of the natural philosophy with which he had dallied just two years earlier. Indeed, he said in it that without a clear vision of the processes of life, natural philosophy wallowed in dismal ambivalence, incapable of really reflecting nature because it was too facile and superficial for simple experiments. He added later, to emphasize the difficulty of research worthy of the name, that the relationship of the scientist with nature necessitated at the same time observation and experimentation. Observation must be simple, patient, honest, and without preconceived ideas. As for experimentation, it was artificial, impatient, laborious, passionate, uncertain, and like a leap into the unknown. This contrast of observation with experimentation, although rather rhetorical, permitted Müller to come back to what was always for him a major preoccupation: the difficulty of carrying out a good experiment and the overriding need to be always able to reproduce an experiment—with the same results.

In the text that followed, he eulogized morphology but also physiology: The experiment was the ferment necessary for the growth of the mind. The field of physiology was not to think about nature in an abstract fashion but to experience it—and then think about it. He concluded by observing that scientific research had something religious about it, that it had its own ceremonies and clerics. He was thinking of Humboldt and Goethe, whom he much admired.

Later when he was a teacher at the University of Bonn in 1826, he published a voluminous work on the comparative physiology of vision,[8] in which he dealt as much with the anatomy of the insect eye as the psychological basis of human eye movements and gaze. A year later, there appeared a second work in the wake of the precedent titled *About Fantastic Visual Phenomena.*[9] It was dedicated to philosophers and physicians and bore on the back cover a citation from Goethe on the various possibilities of behavior in darkness. In it Müller took up and developed his famous law of specific nervous energy, which we examine in detail later. It stated, notably, that the eye and the ear could only produce, respectively, visual and auditory sensations. Each sensory organ responded in its own fashion whatever the external stimulus. A blow on the skin was perceived as pressure exerted on it, but the same blow to the eye provoked a sensation of light, like looking at a starry sky. This specificity led Müller to believe that these perceptions were triggered in the sensory organ and its nerve, rather than in the brain, a point of view that was vigorously disputed by his contemporary, Rudolf Hermann Lotze.[10] So, Müller analyzed systematically, including some experiments on himself, various "fantastic" visual experiences, those of dreams, hallucinations, religious ecstasy, and diabolical or magic apparitions, all independent of any light stimulation and appearing even in total obscurity, but strongly dependent on the psychocultural background of the subject and what later we would call the unconscious or subconscious.

Müller stayed in Bonn until 1833 and was then appointed to the chair of anatomy and physiology in Berlin to replace his mentor Rudolphi who had died. In the meantime, he had conducted a great deal of research on invertebrates and had begun writing his monumental *Handbook of Human Physiology*,[11] in which he summarized the biological knowledge of the time, from anatomy to chemistry and physiology, from plants to animals and man, concentrating particularly on the problems of the organism and animal life. In Berlin he continued his career in research and was very active in comparative invertebrate anatomy, but he was especially involved as a teacher and discoverer of men. During his academic life, he benefited from the friendship and permanent help of Alexander von Humbolt, the very influential councilor to the king.

He was a tireless traveler and, toward the end of his life, was involved in a terrible shipwreck off Christiansand, from which he was rescued just in time and which left him very traumatized. Intractable insomnia and recurrent depression undermined his health, and yet he worked so hard that a family friend called him "man of iron, fed on bronze."[12] During one of his crises in 1858, Müller died prematurely as a result of, it seems, a large dose of morphine taken voluntarily or not. His friend Humboldt, age 90, pronounced a moving farewell eulogy at his graveside.

Nemo psychologus, nisi physiologus

"No one can be a psychologist without being a physiologist," said Müller in his doctoral dissertation in 1822. Four years later, strongly influenced by Spinoza's *Ethics*, he added that the results of all physiological research must be, in the end, psychological in nature, for the mind was only one particular form of life among others.[13] An astonishing fact is that he even dared speak of will and consciousness in animals, which was still scientific heresy for many even 100 years later. But in a speech in 1858 on the occasion of Müller's death, another of his students, Virchov, proposed a rather different interpretation, somewhat removed from Spinoza's and taking into account later evolution of the scientific concepts of his master:

that only a physiologist could become a psychologist and that no philosophy could emerge starting with nature.[14]

In his study on the relations between Schelling's natural philosophy and Müller's audacious physiology, Tsouyopoulos[15] noted that both disciplines faced an identical problem at the beginning of the nineteenth century: Life had become the central problem of monist anthropology. Indeed, the dualist model of two heterogeneous substances, a soul controlling a body machine, hardly stood up to the empirical results of contemporary physiology that demonstrated the complexity of vital functions, such as respiration, digestion, reproduction, and excitability. The phenomena of regeneration observed in polyps were equally difficult to make compatible with the "man machine" of Julien Offray de La Mettrie. The monist model was favored by the philosophers, thanks to Leibniz's *Monadology* and Kant's critical work. Reactions to this monist vision of life were many and diverse, but it is certain that the numerous tendencies that ran through romantic medicine in Germany were at the time largely dominated by the person of Schelling. In this world searching for a monist model, Müller had to forge a conviction and from it derive a rigorous scientific method. He was also the man of transition between romanticism and physicalist physiology.[16]

What was this method capable of making a science of physiology? After rejecting from the start as ineffective both hard-line empiricism and its opposite, a mystical or speculative approach, he adopted a voluntarily polemic position, rejecting pseudomethods such as excessive rationality or false natural philosophy, and practicing intelligent physiology.[17]

Avoiding excessive rationality in the practice of physiology and medicine: Life science was a doctrine built from numerous empirical elements of knowledge and therefore much more than a simple logical link between a large number of empirically observed facts ordered according to Kantian categories.[18] Röschlaub,[19] professor of clinical medicine at Bonn, also said approximately the same thing in observing that at the patient's bedside rationality alone often did not permit the unraveling of the complex tangle of multiple causes involved in disease. Further, let us recall here the law of specific sensory energy mentioned earlier: A blow on the skin produced a sensation of pressure while a similar blow on the eye gave the impression of seeing stars. This observation alone was perhaps sufficient to show that the law of causality, so dear to Kant and physicists, might not apply simplistically in life, in any case not without a better understanding of the living being, which was precisely the task of the physiologist.

Rejecting false natural philosophy that mixed empirical and speculative facts: Here Müller targeted a concept of natural philosophy that deviated from Schelling's tradition and that considered life not as a process of perpetual self-reproduction but as an ensemble of functions analogous to natural phenomena, such as described by chemistry, physics, magnetism, or in terms of polarity.[20] Müller was therefore not condemning natural philosophy as such but rather its deviations that involved arbitrary and sterile analogies instead of searching known physicochemical mechanisms at the very heart of life. In fact one must recognize with historical hindsight that "true" natural philosophy doubtless contributed to giving a more complex and, in a way, more realistic vision of the living organism than German pragmatic medical science of that time could hope to obtain.[21] In support of this paradox was Schelling's aphorism that matter was a single entity made up equally by unity in diversity and diversity in unity. These two aspects made up the whole of matter, and each expressed the whole of its substance.[22]

Practicing intelligent physiology: Here Müller insisted on his concept of observation prior to experimentation, a vision born of personal reflection in which the perception of the scientist, far from being passive, resembled more that of the artist.[23] He shared this concept, far removed from that of Kant, with Schelling's "intellectual vision" and even with Goethe.[24] Experimentation should seek the causes of phenomena. Although this approach was often practicable in the physicochemical domain, it was not always possible in the living world, where a simple stimulus could produce complex effects such as glandular secretion or sensory perception. In this case, said Müller, one needed to imagine an experiment that fitted the degree of complexity of the observed facts, for example, looking for the effect of suppressing a particular function of the organism, knowing that this type of experiment always gave concrete and unequivocal results. It was, however, clear to him that such an experiment—for example, why an animal died after section of both vagus nerves—took him away from a mechanistic explanation of life as much as from a metaphysical one.

Müller and Goethe

Young Johannes Müller admired in Goethe the man of science as much as the poet. Already, his *Habilitation* thesis bore witness to Goethe's influence on his concept of scientific activity, which had to be guided by rigorous observation, resorting to experimentation only with prudence.

Furthermore, according to du Bois-Reymond, it was obvious that even in the organization and manner of presenting his writing he found an inspirational model in Goethe's color theory. What is more, his wife, who was a professional singer, adored the poet, and he knew many of his poems by heart.[25] However, Müller always preserved his scientific judgment and did not hesitate to criticize important aspects of Goethe's theory of colors. Even if he accepted the basic premise of the theory, he nevertheless thought it indispensable to clarify a number of aspects, such as the phenomena of refraction that he had observed. He would have liked a more systematic study and pleaded for a mathematical analysis that might not have revealed a deep significance for color vision, but he would have doubtless spared the author of the color theory many hostile reactions and objections.

Müller was convinced that the color theory was a good starting point for the physiologist. It could hardly be criticized at the level of subjective observation of the phenomena because they depended on the nature of sensory organs. Nothing else enabled one to know what happened within them. It was indispensable to pursue this analysis, he said, for that alone could provide proof that our visual representation of the outside world allowed us or not to recognize and legitimize a so-called external reality. But the analysis of the conditions for the appearance of colors was of crucial importance: Essential in this respect were physiological observations of the structure and function of the retina.[26] Later, without openly criticizing Goethe's theory, which derived colors from a "dynamic" interaction of light and dark, he nevertheless found unconvincing the attribution of any active role by parts of the retina plunged in the dark because they were at a "zero level" of activity.

When he became a professor, Müller wanted to meet Goethe and have a discussion with him.[27] With much precaution, because he was young and still little known, he wrote to the poet, then age 77 and at the summit of his fame, and requested a meeting. He enclosed with his letter a copy of his publications about which he asked his opinion. This letter was very long and heavy with respectful praise for its addressee and detailed explanations of his research in Bonn, but it carefully avoided raising problems that could have displeased the master. For example, he only just touched on the role that physics might have usefully played in the study of sensory organs, other than visual ones.

Goethe's reply came a month later, courteous but rather cold. He thanked the young professor for having sent him his works, but he seemed to have been much more affected in reading those aspects that disagreed with his own views than those that were firmly in line with his. For example, he spoke of diverging paths, inevitable between scientists with different premises, and that it was useless to try to correct them by no matter what agreement. It is significant that at the same time he wrote to Jan Evangelista Purkinje (1787–1869), the great Czech physiologist, who had also just sent him a work on vision. It was a most warm letter expressing his friendship and his total approval of his research, accusing the ghost of Newton of still exercising the same domination as the devils and witches during the darkest centuries of history.

Still not discouraged, Müller had the occasion to go to Weimar two years later, where Goethe received him. But it was only 12 years later in his *Treaty of Physiology* that he disclosed partially, and only indirectly, the details of their meeting. He described the disagreement that separated them about luminous images that appeared to them both when they were in bed, not yet asleep but with their eyes closed. For Müller these images were never symmetrical, and he could not make them appear or change them voluntarily. Goethe, on the contrary, could decide at will the subject of his images, which then appeared without further voluntary intervention and changed continuously and symmetrically. Müller concluded that they each had their own faculty for ideation, and that their natures were clearly very different, Goethe's having the full power of a poet, whereas his own tended toward a search for reality and what happened in nature. What is more, it is amazing that in the *Conversations of Goethe with Eckermann*,[28] which related on an almost daily basis Goethe's preoccupations at that time, the chronicler made no reference to this meeting (October 10, 1828, at midday), but on the contrary related in great detail a mundane lunch given the same day by the master to some visiting friends—the poet Tieck and his family—immediately after the departure of the young professor.

The disappointing relations between the spirited physiologist and the glorious but too self-satisfied old man could be explained, according to Koller, by the precocious and progressive distancing of the physiologist that the poet-scientist must have clearly noticed. But there might also have been wounded self-esteem, justifying Friedrich Nietzsche's words: "Is not wounded vanity the mother of all tragedies?"[29]

Science for Science's Sake

The laboratory of anatomy and physiology that Müller led in Berlin beginning in 1833, and where young Helmholtz had just been appointed, was representative of the concept held by German universities and medical faculties at that time. By creating the University of Berlin in 1809, Wilhelm von Humboldt had given lasting impetus to a curriculum that respected academic freedom and in parallel encouraged the practice of scientific research. But what research? Certainly not research in the service of teaching or with practical applications, even in medicine. Indeed in 1840, teaching of a technical nature was still supposed to be sufficient for the needs of the state. Applied research in chemistry, for instance, was carried out in the "factory laboratory" of a friend of Müller, Justus von Liebig,[30] for purely utilitarian ends. University research was rather "research for research's sake," to push back the frontiers of knowledge, to select the most talented students for an academic career, and finally to allow the state authorities to promote the best researchers in the university hierarchy.[31]

The laboratories, constructed between 1820 and 1840, often had rather cramped accommodation, where only the best students had the right to enter and the technical equipment was still blatantly poor. Only after 1840 did researchers of the generation of Helmholtz, du Bois-Reymond, Brücke, and Virchov undertake to institutionalize laboratories of a new type that would integrate research with theoretical and clinical training of future physicians and permit the average student to train for the scientific practice of his profession.

Such a spectacular evolution in the conception of university research laboratories was not to be seen in the context of the normal development of good laboratories but was rather explicable by a truly ideological rupture affecting the whole of German society.[32] After the Napoleonic wars of the beginning of the century, an intellectual bourgeoisie had emerged, to which belonged the fathers of Helmholtz and du Bois-Reymond. Both dreamed of a united Germany and subscribed to a moderate liberalism on both the political and social planes. Helmholtz's father, the great friend of Fichte, totally subscribed to the latter's philosophy that called for a liberal republic, the promotion of empirical objectivity in science, independent of the subject, and the rejection of Hegel and the sectarian philosophers of "identity." But it equally rejected materialism as dangerous for the spiritual serenity necessary for the exercise of philosophy. As to du Bois-Reymond's father, of very modest origins, he liked to recall his social ascension and how, from a poor "beggar from Neuchâtel," he had become a civil servant in the Prussian administration. He had published privately and under a pseudonym a voluminous work devoted to social problems in Germany, in which he considered that industrialization engendered poverty and that it was the way in which economic competition destroyed the bonds of solidarity and even religious values.[33]

Before 1840, the university, and especially its research laboratories, fitted well this vision of the bourgeoisie that expected it to educate a moderate liberal middle class that would one day achieve the unification of Germany in this spirit. By his civic spirit and scientific prestige, Müller was the typical example of this liberal bourgeoisie, which procured for him a high esteem on the part of his contemporaries and the rulers of the country. He published numerous scientific works of high quality. He fought for an empirico-rational study of life, which he hoped would lead to a better philosophical understanding of it. His science did not aspire to exert an operational role in nature. He devoted most of his time to training elite academic scientists in the respect of civic and spiritual values. On the other hand, one might mention his colleague Johann Lucas Schönlein, Helmholtz's former clinical professor. He was a brilliant teacher and known as an outstanding physician, having introduced to diagnostic practice in Berlin a reliance on physiological and pathological causalities, and he had no less objective merit than Müller. But it was considered that he had not published enough, that his knowledge was only accessible to a limited circle of students, and that his civic spirit was not sufficiently developed.

After 1840, a truly ideological revolution began that led, on the political level, to the insurrection of 1848, but above all to the modernization of Germany and, what interests us here, of its research laboratories. The desire to unify the country produced the idea of a customs union between the states that would one day form Germany. Railways, roads, and canals were built to link different states. Free trade of goods inside the future Germany replaced old customs barriers, but the latter were still maintained with other European states. Industrialization was undertaken enthusiastically in an attempt to compensate for a considerable delay, notably in relation to England.

In the wake, university laboratories modified their policies and adapted their activities to the need to train a greater number of senior personnel. For example, high-level physical and chemical research was henceforth undertaken essentially at the university. These modifications were also felt in physiology laboratories. Müller's young collaborators were the representatives of this new wave and were ready to kick over the traces in order to have access to the most sophisticated technical equipment to carry out their research, the results of which in fact were of benefit to teaching and to society in general. Physiology became more and more "physical" and medicine a much more scientific and structured practice. In fact it was high time because until then medicine underwent no quality control by the state. Consequently, there were too many practitioners forced to sell their services on the cheap, attract patients by publicity pamphlets, or employ the services of healers and charlatans to reinforce their popularity. Apothecaries and heterodox practitioners such as chiropractors and homeopaths were particularly worrying. Certain people thought that the law of the market, based on the efficacy of the treatment, would suffice to select some physicians and eliminate the rest. But for others only a health policy organized by the state in collaboration with the faculties of medicine and their laboratories would be capable of guaranteeing the quality of medical care.

Helmholtz, du Bois-Reymond, and their friends were obviously very much implicated in this inevitable ideological and scientific revolution. They were the coactors. Their opinions, both philosophical and political, were quite different from their fathers' and their master Müller's, all three of whom were moderate liberals opposed to materialism. Although liberals themselves, the younger men were tuned to their own era and, as we see later, became more or less convinced or confessed adepts of materialism, which was something quite different.

Thus, at this key turning point between the old and the new style of laboratory research, Hermann Helmholtz finished his studies and was preparing his thesis for the doctorate in medicine, for which Müller had given him the project of an anatomical study of nerve fibers in some invertebrates he had in his collection. Scarcely had he begun his work when he was struck down by typhus and hospitalized in a serious state at the university hospital of La Charité. He stayed there several weeks before recovering, and he then began a long convalescence. As soon as he was out of the hospital, he rushed to buy, with the capital he had saved during his illness, a medium-sized microscope to enable him to advance his research more quickly.

A few months later, he wrote to his father to tell him that he had submitted his first results to his chief, who seemed interested in them. In fact, he had demonstrated for the first time that nerve fibers visible in the microscope in his specimens arose from ganglion cells that had been observed a few years previously, some distance away from these fibers, by the physiologist Christian Gottfried Ehrenberg, who had been satisfied to simply *suppose* that these neurons could be the origin of the fibers in question. Müller advised him to continue his work and add to his observations on three or four species studies on a few more in order to make his conclusions even more convincing. Later in his letter, Hermann begged his parents' forgiveness for the inevitable delay in the date of the award of his doctorate and the disappointment that this would doubtless cause them. He asked them to understand the importance of what was at stake for him, saying that he was ready to give up his holidays for this work. His parents obviously did not oppose their son's project, and Müller, clearly well pleased with him, gave him permission—a significant favor—to use his own research microscope.[34] Encouraged by this support, he pursued his work and managed to demonstrate that, in all the species he had been able to observe, the peripheral nerve fibers arose in continuity from the ganglion cells described by Ehrenberg, and that therefore these neurons were of primordial importance in the development of the nervous system of the animal. Helmholtz defended his thesis in public. Its title was *Nerve Fibers Arise from the Ganglion Cells Discovered in 1836*, and he received his doctorate in medicine on November 2, 1842.

4 Vitalism: The Best and the Worst of Things

Life is the ensemble of functions that resist death.
—Xavier Bichat (1771–1802)[1]

The doctorate that young Hermann Helmholtz had just obtained with distinction at the age of 21 by no means marked the end of his studies. There remained a year of internship at La Charité, which obliged him to spend a few weeks in each important service of the university hospital, after which he would need to start his military service and pass the state examination that would qualify him as a physician.

When he finished his internship in September 1843, he was posted to the garrison at Potsdam, where he prepared his final examination that he passed two years later but only moderately well. He was not accepted as a surgeon as he had hoped but as a physician and traumatologist. His enthusiasm for practical medicine was probably rather lukewarm, and his heart was on other things. Indeed while still in Berlin in his letters to his parents at the beginning of his internship, he complained of not having enough free time for his own work. When a few months later he found a little more freedom, he immediately returned to Müller's laboratory.[2]

Helmholtz and the Shadow of Theodor Schwann

In the laboratory, Helmholtz found his seniors, du Bois-Reymond and Brücke, who helped him with their advice and explained the various problems the research world was facing at the time. Thanks to them, he quickly became aware of the unusual atmosphere that reigned in the laboratory between the assistants and their leader. There was an undeniable generation gap between them, as we discussed briefly in chapter 3, and that was not surprising in itself. What was surprising was the cohesion and genuine efficacy of the group, explicable as much by Müller's personality, which encouraged excellence in research training above all other considerations, as by the assistants' respect for their leader, whose skill, honesty, and modesty they admired. However, on several occasions, there had appeared dangerous fissures in the fine edifice that Müller was building for the future, notably when he arrived in Berlin from Bonn.

Müller had appointed two assistants with whom he pursued his research with much enthusiasm, Jakob Henle and Theodor Schwann.[3] Later in 1858, the latter wrote to du Bois-Reymond how much he admired his mentor for his loyalty and the clarity of his teaching, for his extremely stimulating influence in research, and for his encouragement to experiment and tirelessly pursue his investigations each time a new idea emerged.[4] Theodor Schwann was a zoologist who made his career from 1838 first in Louvain and then in Liège, and he was a great scientist. In Müller's service, he pronounced his cell theory, so founding modern histology, but he was also known for his description of the protoplasmic layers forming the myelin sheath that protected nerve fibers, of which the cells of origin are known as Schwann cells to this day. In 1837, at the same time as Charles Cagniard de Latour[5] and Friedrich Kützing,[6] well before Louis Pasteur, he discovered the basic mechanism of fermentation, describing it as the result of the decomposition of sugar by a fungus—yeast—that derived from it the material necessary for its growth and reproduction. This concept had been vigorously criticized by the chemist Liebig, who saw it as vitalist and thought personally that yeast was itself decomposing, and that it caused the transformation of fermentable elements by contact.[7] This controversy in fact served as a springboard for Helmholtz's first experiments described later.

The importance of Schwann's cell theory was that cells of all tissues possessed a nucleus in their protoplasm, and for the first time the structural unity of cells, animal as well as vegetable, emerged as a general principle for the formation of organic bodies. Cells were organized according to defined rules, he stated. In his later work, he built the bases of embryology by observing that the egg was a single cell that developed into a complete organism. Another of Müller's students, the pathologist Rudolph Virchov, completed Schwann's work by proclaiming in 1858 the axiom *Omnis cellula e cellula*, meaning that all cells were derived from the division of an earlier cell. He generalized his theory to cancer, in which the multiplication of abnormal cells necessarily originated from a single cell that had become cancerous.

It seems that Schwann was the first of Müller's assistants to discuss some of the latter's supposedly vitalist concepts, and even oppose certain

of his teleological concepts with a physicalist concept of life, so entering
into the confrontation between vitalists and antivitalists that was so
heated between 1835 and 1850. Speaking of muscular contraction, he
said that, as far as he knew, for the first time a vital phenomenon would
be subject to mathematical laws expressed in numerical terms.[8] We see
next what we might think of Müller's vitalism, but what is important for
our account is that Schwann had two ideological heirs, du Bois-Reymond
and Brücke, and they kept young Helmholtz informed about all these
discussions.

Was Müller Really a Vitalist?

Most authors rightly consider that, after his stay in Berlin with Rudolphi,
young Müller had definitively rejected natural philosophy as a theory of
reference for his physiology. Nevertheless the label of vitalist remained
indelibly associated with him partly as a result of statements by his students, perhaps repeated rather hastily by certain recent historians,
doubtless keen to make the great scientist Helmholtz rise from the
obscurantist physiology of the early nineteenth century and to credit him
with the same glory as Perseus freeing Andromeda from the sea
monsters.

But what really was vitalism? To understand the phenomenon, the
reader must forget the discoveries of the last two centuries in the life
sciences and try to imagine the difficulties of those who were interested
in the study of biology during the eighteenth century, trying to find an
explanation for the seemingly definitive coherence and unity of the
activities of men and animals. Among other things, they observed that
living matter was subject to spontaneous decomposition after death.
Further, organic tissues could be reduced experimentally to inorganic
components, such as metals and salts of sulfur or nitrogen, by chemical
and physical agents such as acids, or simply by heating: It was, however,
impossible to manufacture living matter from these elements. So, the
destruction of living matter was easy, but its reconstitution was
impossible.

There were other subjects of astonishment and perplexity. The regeneration of damaged organic tissues, the discovery by the Swiss zoologist
Albrecht von Haller of "irritability" or contractility of muscle and, even
more extraordinary, the observation by the latter that after the death of
an animal its intestine continued to manifest peristaltic contractions and
that the same applied to isolated fragments of intestine as if each had

been able to preserve a fragment of life.[9] As the laws of physicochemical causality were hardly useful to explain these phenomena, it seemed quite logical to invoke an unknown vital force that unified and finalized, and that emerged from the sum of the parts to control the whole organism, a sort of "dominant monad" as Leibniz had suggested.

Vitalism was one of the answers to this quest for a vital principle. It was born in the eighteenth century as a reaction to Descartes and his disciples, who had essentially rejected their predecessors, from Aristotle to Van Helmont, who all recognized the notion of life, without contrasting it to a physical force that dealt only with the inanimate. After all, was not the whole of nature alive? Before Descartes, the soul conferred a certain autonomy on living beings, but it was not destined to oppose natural laws that undermined life from the inside.[10]

The appearance of mechanistic concepts of the living being, so dear to Cartesianism, left the body passively exposed to the aggressions of the environment and to the rigors of physical laws, without the compensation of an active principle that could oppose the successive degradations that led to death and corruption. Hence, the interest of numerous physiologists in what Leibniz called "active dynamics," with living matter finding the source of its activity within itself, a mechanistic explanation of life being possible thanks to a particular "force" intrinsic to matter.[11] In fact, Leibniz stated that all organisms were mechanisms, but not all mechanisms were organisms, which left room for defining the specificity of living things.[12]

Taking up the concept of an "organized" being, without using the word "living,"[13] Kant stated that such a being was not simply a machine because a machine only possessed motive force, but the organized being possessed a forming force (*bildende Kraft*) that it communicated to materials that did not possess it. An organic body was not only organized, it was auto-organized. The physician Carl Friedrich Kielmeyer[14] pursued the same idea, insisting on the effects of circular causality. Every organ underwent modifications, he said, which were so much a function of those undergone by its neighbors that each seemed to be the cause and the result of the causes. It obviously remained to know the nature of what Leibniz called active dynamics or intrinsic force and what Kant saw as the energy that formed and organized matter.

The concept of vital force, which emerged under different names, was first interpreted by Georg Ernst Stahl,[15] the personal physician of Friedrich Wilhelm of Prussia and contemporary of Leibniz, with whom he had several stormy discussions. Best known for his definition of "phlogisitic," he attempted to derive an autonomous chemistry from the obscurity of alchemy, which was only achieved later by Antoine Lavoisier[16] at the end of the eighteenth century. Stahl claimed that life was due to a specific principle that fought against spontaneous corruption because the only thing that physics or chemistry could explain was the corruption of the body after death. This principle was of a metaphysical order: a thinking soul, doubtless situated in the brain, acting via the nerves to maintain or repair the body and combat its inevitable corruption. This soul was capable of resisting the course of physical laws and so differed completely from the pre-Cartesian soul, which was not able to do so.

However, this animist vitalism of Stahl was opposed forcibly by those who refuted as metaphysical dross all interference of a soul in the play of physical laws. They could not accept that a living being could ever be incapable of ensuring the control of its own vital processes. This opposition was led essentially by the French, such as Paul-Joseph Barthez,[17] Théophile de Bordeu,[18] and, above all, the great theoretician of "life," Xavier Bichat,[19] who bequeathed to us his famous definition: "Life is the ensemble of functions that resist death."[20] However, the nature of this vital principle was not really defined even by Bichat, who found it unnecessary to explain it. It consisted of a principle that was distinct from both the thinking soul and anything known in eighteenth-century physics. Bichat hinted that it might be useful to take inspiration from Newton's models that did not define the earth's gravity but limited themselves to observation of its effects. Therefore, Bichat favored giving priority to observation and experiment rather than searching for causes and explanatory principles.[21]

Bichat's vitalism thus had nothing to do with metaphysics and limited itself to concluding that, in the state of knowledge at that time—and not a priori—it was impossible to reduce the living being to a machine explicable in physicochemical terms. If we make a leap forward to the more recent past, we note in fact that, with the exception of certain vitalists of the Montpellier school who lost themselves in vain spiritualism, even scientists of the class of Claude Bernard remained confronted by the same methodological problem of vitalism. On the one hand, he described "vital" as the "organic properties that we have not yet been able to reduce to physicochemical considerations, but there is no doubt that we shall succeed one day." On the other hand, he said: "When a chick develops in an egg, it is not the formation of the animal body as a group of chemical elements that essentially characterizes vital force. This group is only formed according to laws that govern physicochemical properties of matter; but what essentially belongs to the domain of life, and does not belong to chemistry or to physics, or to anything else, is the guiding force of this vital evolution … for its whole duration the living being remains under the influence of this same creative vital force, and death looms when this can no longer be."[22]

If these statements by an uncompromising scientist like Claude Bernard seem strange and even ambiguous, today they were not unreasonable if we take into account the context of the period in which they were pronounced. According to Claude Debru,[23] it was just at the moment when the vital was closest to the physicochemical that its irreducibility appeared most obviously, and antireductionist statements were opposed by reductionist statements in Bernard's thoughts. To the "we can reduce" there succeeded a realist "we cannot invent," thus opposing the vital syntax to the physicochemical alphabet. It was in fact only in 1901 that Franz Hofmeister[24] demonstrated the reversibility of the action of certain cellular enzymes that contributed to the construction, and not only the destruction, of organic matter, thus announcing the end of chemical vitalism. As to the "guiding force of this vital evolution" described by Claude Bernard, it had to wait even longer before receiving an exclusively physicochemical explanatory hypothesis thanks to the discovery of the genetic code in conjunction with Darwinian selection.

So what was the situation of Johannes Müller's "vitalism" about which his young collaborators were so reproachful? In the *Prolegomena* at the beginning of his *Elements of* Physiology,[25] he posed the problem from the outset whether fundamental forces that underlay, respectively, organic and inorganic matter were different or whether the forces of organic life were merely modifications of physical and chemical forces. Having reviewed the chemical mechanisms related to organic life known at that time and having noted, like most of his contemporaries, that life was in opposition to death and putrefaction, he took great care not to take sides and neither affirmed nor denied the possibility of one day realizing the synthesis of organic matter, especially as urea had just been synthesized in the laboratory by Friedrich Wöhler[26] in Berlin (1828) "without a kidney or an animal."

Müller's attitude was doubtless dictated by his intelligence as much as by his scientific objectivity, but a sprinkling of political prudence may well also have been involved. Some years later, the materialist Jacobus Moleschott was relieved of his university duties because he stated that thinking led to the appearance of phosphorus in the brain. It is, however, significant that Müller's collaborator, du Bois-Reymond, the founder of electrophysiology, always critical toward his mentor, should write to a friend that it was uniquely due to Müller that he had returned to physics.[27] Müller was also criticized for his concept of specific energy of the eye, which resulted in it being able to see but not hear. But such criticism had little foundation for anyone who reads his *Physiology of Vision* and realizes how rational the author was and how rigorously he defined his terms. The word "energy," for example, related to Aristotle's concept of energy and contained nothing that evoked any vital force because he used it to mean "sensation" and nothing else. Ernst Mayer[28] recalled a citation by Müller stating that *Lebenskraft* (vital force) acted in all organs as supreme cause and effector of all phenomena according to a defined plan. He pointed out that the qualities which Müller attributed to vital force were strangely close to our modern concept of a genetic program:

1. It was not localized in a specific organ.

2. It was divisible into a large number of elements, each expressing the properties of all the rest.

3. It disappeared with death, leaving nothing behind.

4. It acted according to a plan.

It was thus telenomic.

What Müller's collaborators reproached him for was doubtless more related to the generation gap, as we said earlier, and their disappointment not to see their own credo in the omnipotence of physics adopted from the start by their mentor. It is useful to read Georges Canguilhem,[29] who described beautifully the extreme complexity of vitalism. A particularly strong statement was: "In the end classic vitalism accepts the insertion of the living being in a physical medium to the laws of which it constitutes an exception. Therein lies, in our opinion, the inexcusable philosophical error."

Let us leave the last word to Helmholtz, who in his memoirs acknowledged: "In these studies I was influenced by a teacher of great depth, the physiologist Johannes Müller" who "struggled with the puzzling questions of the nature of life between ancient, essentially metaphysical, concepts and newly developing concepts of the natural sciences, but he was more and more convinced that nothing could replace knowledge of the facts; that he himself was still struggling perhaps made his influence on his students even greater."[30]

Putrefaction and Fermentation: Half Failure or Half Success for Helmholtz?

Schwann discovered that, in the presence of yeast, grapes produced wine and hops beer; the interpretation of this phenomenon—the proliferation of yeast that fed on sugar and secreted alcohol—had been considered as vitalist by Liebig, for whom fermentation would thus be simply a prolongation of biological functions. Liebig's opinion was that the proliferation of the yeast was a consequence of fermentation and not its cause, which was in fact, like putrefaction, an exclusively chemical process.

Helmholtz saw in this controversy, which extended far beyond the doors of the laboratory in countries with an important brewing industry, a chance to act as arbiter. So he conceived of a series of experiments in which, for the first time, he abandoned microscopy for chemistry and carried out in Müller's laboratory and later in that of his physics professor, Heinrich Gustav Magnus,[31] who enjoyed important technical resources and welcomed him warmly. But there was also the burning controversy about vitalism, in which he was impatient to participate and even do battle.

In his memoirs,[32] Helmholtz recalled how most physiologists of his time were adepts of Stahl, who explained the decomposition of the body after death by the fact that the soul or vital force ceased to control the physical and chemical forces of the organism, which, left to themselves, produced putrefaction. "In this explanation, I suspected something contrary to nature, but it took me much effort to translate this suspicion into a definite question. In the end, in my last year of studies, I found that Stahl's theory gave to all living bodies the nature of perpetual motion." Helmholtz also recalled that he had already discussed the problem of perpetual motion a lot when he was still living with his father, and that later he had studied its physical and mathematical aspects in depth by reading Bernoulli and d'Alembert. A "good" question would have been: "What relationships must there be between the various forces of nature for perpetual motion to be possible?" But the question he actually asked was: "What relationships must there be between the various forces of nature for no overall perpetual motion to be possible?" A further question was: "Do all these relationships really exist?"

We see later how Helmholtz was drawn into writing his famous treatise *On the Conservation of Force*. But first he wanted to test some of his ideas experimentally. From this stemmed his interest in the problem of putrefaction that seemed to him to be at the heart of the vitalist controversy.

So he conceived an experimental apparatus in which he could isolate putrefactional activity in unicellular microorganisms, such as found in rotting matter and among which were the yeasts discovered by Schwann that he believed to be the cause of alcoholic fermentation. He distinguished them from gaseous or liquid vectors of putrefaction, which for Liebig were the exclusively chemical cause of putrefaction and fermentation. Helmholtz placed pieces of meat in a bladder of which he carefully sealed the orifices so as to prevent any passage of microorganisms, but leaving it possible for gases or liquids to pass. After plunging this hermetically sealed bladder in a liquid that was putrefying, he observed the passage of this liquid by "endosmosis" across the bladder wall causing the putrefaction of the meat, but that it was very slow and incomplete.

From these results he concluded that:

Putrefaction could take place independent of life because it had taken place, at least partially, without the intervention of micro-organisms, kept out of the bladder.
Putrefaction of the liquid in the bladder was not however complete without contact with the external liquid containing micro-organisms and that putrefaction so modified by organisms was in fact fermentation. Putrefaction clearly resembled a vital process by virtue of its taking place in the same matter, and its power of propagation.[33]

Helmholtz must have found his results a little disappointing because he had not succeeded in excluding either of two possible causes of putrefaction and certainly he had not refuted Liebig's ideas. He had clarified the facts of the problem by stating, with experimental data, that fermentation or putrefaction depended respectively on the presence or absence of microorganisms. Practically, he had not succeeded in his secret desire of finding arguments against the hypothesis of vitalism; on the contrary, he had given the impression to his contemporaries to have strengthened its hold. The end of this story was written in 1857, when Pasteur discovered that different types of yeast, and not different fermentable media, produced a specific type of fermentation. The yeast was thus the cause. Fermentation, like putrefaction, was determined by the action of living cells and not simply reducible to spontaneous chemical processes associated with death.[34]

However, the most important thing was that Helmholtz had demonstrated that he could carry out experiments, and he now better grasped their requirements and their limits with intelligence and honesty. Soon after, he was to renew his struggle against vitalism, with more success this time.

Frogs, "Those Old Martyrs of Science"[35]

Not at all taken aback, Helmholtz pursued his relentless crusade against Stahl's vitalism. Rejecting more and more the hypothetical intervention of metaphysical forces in organic matter that would transform it into an automaton with a sort of perpetual motion, he realized that if he wanted to progress along his chosen path, he would have to test on an animal a simple experimental paradigm that was likely to give quantifiable results. What could be simpler than a frog's muscle, isolated but alive and still capable of contracting if excited by an electrical current? Any mechanical or thermal energy resulting from the contraction would obviously come from the metabolism of the muscle, that is, from energy reserves in the actual tissue, and it would be possible to confirm it quantitatively. Were the mechanical force and heat produced in an organized body derived entirely from its metabolism?[36]

This was an ambitious project that led him to study successively chemical modifications of the muscle caused by its contraction, as well as the heat produced by this activity. The project was all the more justifiable because, although it was known since Lavoisier that a man at work needed more oxygen than a man at rest, and that muscular work consumed substances that might be measurable and were replenished after the effort, there was total ignorance of the nature of the substances consumed and where these metabolic processes took place.

So the first step was the chemical metabolism of the muscle during contraction. Helmholtz dissected the paired thigh muscles of "those old martyrs of science," which he separated from the body of the frog and sectioned at the level of the ankle. He submitted the first thigh to 4,500 electrical stimulations in quick succession and provoked vigorous muscular contractions until the muscles were exhausted and responded no longer. The second thigh, which had not been stimulated, retained its excitability and served as a control for comparative metabolic measurements.

Methods for extraction and chemical analysis of organic materials were still rudimentary at that time. Nevertheless, Helmholtz managed to find significant differences between stimulated muscles and those that had remained at rest. In muscles that were exhausted by repeated contractions, extracts in distilled water contained a little less albumin than resting ones. On the contrary, the ratios were reversed after extraction with alcohol, with active muscles containing more chemicals than inactive ones. Helmholtz regretted bitterly that the poor state of the available techniques for chemical analysis prevented him from proceeding any further without identifying the substances involved in muscular contraction, and thus unable to formally demonstrate the relations between metabolism and muscle activity. He claimed, however, not unjustifiably, to have proved that a chemical change in components of the muscles had indeed taken place during their activity.[37]

What did not emerge from the necessarily rather abrupt report of these results were the great strides forward made by young Helmholtz in the elaboration of his experimental methodology. First, his systematic comparison of extracts of active muscles with those of homologous inactive ones permitted a very elegant demonstration of specific effects of the contraction of a muscle on its metabolism, without needing to take into account possible unknown external influences, because they would have been identical for the two muscle groups. Further, he demonstrated, as a good mathematician should, that repeating the experiments allowed him to determine a mean value for the results with greater precision the more the experiment was repeated.

Then followed the second stage of his experimental project: heat production by muscular contraction. He had realized more and more clearly that the ideal of observation inherent to anatomy was slowly giving way to an experimental physiological approach from a physical and chemical standpoint. He had therefore prepared himself carefully for the second series of investigations, particularly as he had learned a great deal from an essay on animal heat that he had been asked to write for a medical encyclopedia.[38] In this extensive work, he had produced a critical revue of data from insects, birds, and mammals. He found them unreliable because of their lack of rigor, such as too few measurements to be useful quantitatively, an uncontrolled environment, and ignorance of physiological variations in temperature.[39] His judgment had been especially harsh for theories dealing with the origins of animal heat, notably that of Liebig, who assumed that it was due to the combustion of elements in the blood by inspired oxygen. Helmholtz considered this theory in error because he was convinced that the heat produced (e.g., by a contracting muscle) came essentially from the combustion of its own elements.

He decided once again to use the isolated frog muscle, an ideal preparation to control experimental variables and suppress any influence, such as heat, coming from the rest of the body. He wrote enthusiastically to his friend du Bois-Reymond that he had finished the construction of the necessary electrical apparatus and that he was waiting impatiently for the spring, and the frogs, to begin his work. When that moment arrived, he proceeded as before, but this time he compared thermogenesis in resting muscles with activated ones, obtaining useful results even from small differences in temperature between the two muscle states. Unlike for his chemical experiments on muscle, the new apparatus was very complicated, sophisticated, and specifically conceived for these experiments. There was a new type of electrical stimulator with an induction coil enabling the production of a continuous contraction (a *tetanus*) thanks to a series of short but intense impulses. There was also a thermometer formed of thermoelectric sensitive electrodes inserted in the muscle being tested that could detect variations in temperature of the order of a thousandth of a degree. What did he observe? The temperature of the tetanized muscle was some 0.2 degrees higher that the control muscle, which could not be explained by the arrival of warmed blood from elsewhere, but rather by the combustion of substances in the muscle itself.[40] So the circle was closed and his initial objective basically achieved because he had enough evidence to describe, at least qualitatively, the unified processes by which an intramuscular chemical provided mechanical force and heat. Indeed this series of experiments was to find strong support in a theoretical paper that he was writing in the secrecy of his office at the hospital, *On the Conservation of Force*, which we discuss in chapter 5.

This work on frog muscle has remained one of the classics of nineteenth-century physiology, not only because of the importance of Helmholtz's results, but also, and above all, because of the novel character of the experimental methodology and the rigor with which he handled the quantitative data. There was also novelty in the construction of more and more complex laboratory apparatus that had become indispensable for the young science of physiology to benefit from the most recent advances in physics and chemistry. The specific needs of each new experiment required much imagination and technical ability. It is significant that Helmholtz devoted much of his published work to the description of technical and methodological aspects of his research. As to vitalism, he rarely spoke of it from then on because he considered that he had shown clearly that it was not necessary to postulate the intervention of metaphysical forces in a living world of which the coherence and dynamism could be explained in terms of physicochemical processes.

Helmholtz in Love

The reader, impressed by the intellectual talent and ambitious energy of young Hermann Helmholtz—he was only 27 in 1847—should not imagine that he was impervious to the temptations of life and the siren call of love. There is no doubt that he enjoyed life. During his studies, he always found time for the various distractions offered by the town of Berlin in the form of theater, music, excursions with friends, and even long walks. He also consecrated much time to reading great authors from Shakespeare to Kant and Goethe, and he played the piano regularly. But we know of no trivial amorous adventures on his part. Certainly, we might object that, despite the interest of their writings, his biographers were most often very conformist flatterers and would have been wary of accusing Helmholtz of passing fantasies or adventures, even had they known about them. However, in the absence of convincing evidence, such suspicion would seem unfounded because he was very demanding of himself in his rather rigid environment and would have been well aware that any indiscretion—even if he had been tempted at times—might have been of the utmost prejudice for the future course of his career.

At the beginning of 1847, Helmholtz fell in love, quite simply, on the occasion of one of the numerous social evenings organized by his friends or his parents, during which he was often invited to play music or participate in a theatrical presentation. During the evening in question, the young doctor was at a reception in Potsdam by a lady of some standing, Julie von Velten, widow of a military physician whose father had been ennobled for saving the life of Friedrich II on the battlefield. She was herself the daughter of the keeper of the art galleries of the philosopher king. She and her two daughters, Betty and Olga, formed a trio of intelligent women, very gifted for music and art in general and perfectly adapted to the demands of the elitist society of Potsdam.

It was for Olga, the younger sister, that Hermann's heart felt love at first sight that would ultimately unite their destinies. According to her sister Betty,[41] "Olga was not beautiful but fine and charming, not vivacious or outstanding, but intelligently attentive and a keen observer; she had a quick mind, was amusing and witty, and sharp to the point of sarcasm. But above all she was surrounded with a veil of femininity and plain, simple purity—something quite irresistible." As for Dr Helmholtz, when he visited the von Velten ladies for the first time, he seemed to Betty somewhat strange, "very serious and introverted, a little clumsy and uneasy among the animated and mundane young men. He fitted perfectly what I had been told when he was introduced: a very intelligent man, but one had first to dig a little deeper: then one discovered a mine of treasures."

He was quickly adopted by the family and felt so at ease at their home that he imagined, as he said, that he was living out a novel. He often made music with Olga, whose pretty voice he accompanied at the piano. He read her poems that he had written for her, and he played in short comedies in which he particularly excelled in comic roles that sometimes bordered on the grotesque—notably one time, according to his biographer,[42] when he was writing the introduction to his *Conservation of Force*. Thus, as his future sister-in-law Betty admitted, "he grew inseparably into our lives and there developed in him and my sister the realization that they belonged together for life." Their engagement took place on March 11, 1847, but the marriage could unfortunately not happen before Hermann was capable of earning his living to support the family, and they had to wait several long years before they were able to wed.

The more hard-headed reader may well find in this anecdote all the ingredients of a rose-water love story or, as people might have thought at the time, a Biedermeier romance. Biedermeier was a fictive character parodying the ignorant bourgeoisie who believed himself highly literate, during the period of the transition (1815–1848) between the peace that followed the Napoleonic wars and Metternich's repression. It was a time when literary salons contributed to the bourgeoisie becoming literate and literature becoming bourgeoisie. It evoked a comfortable and plush lifestyle but not without a certain nationalism. However, we must realize that, even with today's more emancipated values, it is risky and perhaps futile for the historian to ridicule an epoch and even more so to judge its moral standards.

As to the characters of our story, Hermann and Olga, we can only obtain an idea of their true feelings for each other in the context of an era they did not choose. Hermann's numerous letters leave little doubt that he was really very much in love—not only from their tone and their style but also from the evidence they provide of numerous examples of mutual attention in everyday life. After their marriage in 1849 until Olga's early death in 1859, he remained deeply in love with his wife and full of respect for her. Despite the birth of two children, Katherina and Richard, and her rapidly failing health, she reciprocated by devoting much time to her husband, as a subject in his experiments on physiological optics, as his secretary for many of his papers, and, above all, as a careful critic of his plans for conferences, helping him express his ideas so as to be understood by his audience. It must not be forgotten that by marrying Olga, Hermann proved his absolute disinterest from the financial point of view. Furthermore, he had not hesitated to welcome his impecunious mother-in-law to his young household. Such a gesture when one is oneself far from rich showed a certain flair.

But there is one final aspect of the personality of Hermann that may seem surprising. At a time when, unlike today, feminism was still far from influencing people's minds, he was consistently anxious to see women at last become involved in science or public affairs. He took a certain pleasure in encouraging a woman when he estimated she was more intelligent than her husband. He was very impressed by the interest of English women for scientific problems, and at a conference in Zurich he could not hide his stupefaction at the spirit of independence of Swiss women who not only attended university courses but who would not marry without first signing a contract with their future husband in which they posed certain conditions, notably to be able to leave their home in the city to live for several weeks each year in the mountains.[43]

Helmholtz's ideas on the role of women in society were broad-minded for that epoch and bore witness to his real foresight. The time is perhaps ripe to recall here that, in relation to women, Goethe had not hesitated to say—admittedly twenty years earlier "Women are silver cups in which we lay golden apples ... The characters of women that I have represented are to their advantage; they are all better than we could encounter in reality."[44]

5   Helmholtz and the Understanding of Nature

(der Weise ...)
Sucht das vertraute Gesetz in des Zufalls grausenden Wundern,
Sucht den ruhenden Pol in der Erscheinungen Flucht.

[(the wise man ...)
seeks a familiar law amid the awesome wonders of chance,
seeks a stable pole amid the flight of phenomena.]
—Friedrich Schiller, *Der Spaziergang,* 1795 (Helmholtz's favorite citation:
"pole" here signifies "substance" that does not change ...[1])

The year 1847 was certainly a good one for Helmholtz. His physiological
experiments had achieved most of their initial aims, had enabled him to
perfect his own experimental method, and had earned him a good repu-
tation for emphasizing the place of physics and chemistry in his work.
What was more, the scientist in him had avoided destroying the man in
him: His need for beauty and culture, and his obvious desire to share his
sentiments, had successfully paved the way to the conquest of the heart
and mind of a young woman with whom he hoped to share his life for
better or for worse.

But the time had come to strike a blow. In his experiments on isolated
muscles, he had demonstrated links among metabolism, force of contraction, and heat production. But because of the insufficient development
at the time of techniques for physicochemical analysis, he had not
managed to discover the nature of these links and thus quantify them.
So he resolved to approach the problem from the other end and elaborate a theoretical model of the physical world in which the combined
forces involved would always remain constant despite their apparent
diversity and anarchy. He thus hoped to compensate, by this theoretical
detour, his "temporary" incapacity to use an experimental approach to
prove his intimate conviction of a direct relationship between the energy

released by muscular metabolism and the amount of mechanical force and heat obtained.

Fortunately, his knowledge of physics and mathematics was far superior to that of most of his physiologist colleagues, which was of the greatest value in the conception of his model. He wrote his treatise *On the Conservation of Force*[2] in a few weeks and presented it to members of the Academy of Physics in Berlin, still in the year 1847. The Academy was founded in 1845 at the instigation of Magnus, professor of physics at the university with whom Helmholtz had carried out some of his recent experiments. There he encountered his friend Emil du Bois-Reymond, one of the founders, who had in fact sponsored his membership. He also met the physiologist Ernst Wilhelm von Brücke, the physicists Rudolf Emanuel Clausius and Gustav Robert Kirchhoff, the physician Rudolf Virchow, and, in particular, the engineer Werner Siemens, who was to play a major role at the end of his career by helping him found his Physicotechnical Institute.

He owed much in writing this treatise to du Bois-Reymond, who often discussed its contents with him, helped him write certain parts of the text that had been done too hastily, and finally expressed his admiration for "a historic document of a grandiose scientific conception valid for all time."[3]

Emil du Bois-Reymond, the Friend

Everything seemed to draw Emil and Hermann, two years his younger, together: a very similar family atavism from which they tried to distance themselves, the same professor of physiology and laboratory head, the same repulsion for the deductive sciences inherited from natural philosophy, and common scientific ideals oriented around the importance of physical science in physiology. In brief, they were two cocks in the same farmyard. Yet they were never rivals or jealous of each other, and inevitable conflicts of interest never took the upper hand. On the contrary, they had great esteem and admiration for each other. They corresponded incessantly to discuss their experimental results (168 letters cataloged[4]), conceived common projects together, helped each other with the never-ending design of laboratory equipment, and remained great friends until the end.

But these two personalities did not hide their differences of opinion: They were both obsessed by philosophy and nourished by it, but sometimes with subtle but important differences. Emil was highly cultured and had devoted much time to the study of philosophers. He obviously studied Aristotle, whose work was a normal part of the medical curriculum and to whom Müller often referred knowing that his students would understand. But he also studied other Greeks, Francis Bacon ("knowledge is power"), Descartes whom he could cite from the original texts, Spinoza and Bruno from whom he borrowed a few verses in notes for an essay on *Darwin and Copernicus*, and, of course, La Mettrie and the Enlightenment philosophers, such as d'Alembert and Diderot. He also much admired Leibniz and his attempt to relate philosophy to the physical and mathematical sciences. For du Bois-Reymond Kant was the last in a series of philosophers who knew perfectly the natural sciences of their time and were personally committed to their progress.

Helmholtz had perhaps not read as many philosophers as his friend, but he was even more interested than he in Kant as a philosopher. His critical philosophy, he said, consisted essentially in looking for the sources and justification of our knowledge, as well as in classifying each science according to the measure of the intellectual means employed.[5] There is no doubt that he found philosophy extremely important for the elaboration of scientific reasoning, in contrast to metaphysics, which was of no interest because it was to philosophy what astrology was to astronomy: "To elaborate metaphysical hypotheses," he mocked, "is like fencing in vain against a mirror."[6] Nevertheless, he would do it himself, although in a good cause, in the introduction to his *Conservation of Force*.

Both men felt a real disgust for the "identity philosophers," notably Schelling but also Hegel and above all Arthur Schopenhauer, who had one day stated peremptorily that "he who denies the existence of vital force denies his own existence, and can glorify himself even more for having reached the highest levels of absurdity."[7]

Goethe found no favor with du Bois-Reymond, for whom German science would have been much further advanced had Goethe not interfered, but it had been able to progress despite him. "Even if it may seem prosaic," he said one day, "it is nonetheless true that instead of going to Court, spending paper money of no value and climbing toward the witches of the fourth dimension, Faust would have been better off to marry the girl, raise her child decently and invent electrical machines or vacuum pumps."[8] This was not exactly Helmholtz's view: He was, of course, like his friend, also violently opposed to most of the scientific ideas of the demigod from Weimar, but he deeply respected the poet and his specific vision of nature. Indeed later he made the distinction between the logical reasoning of the scientist and the poetic reasoning of the poet, each of which had its own methodology and different objectives.

Another difference between the two physiologists was their attitude toward the problem of nativism and empiricism, better known today as that of the innate and the acquired. Helmholtz was opposed to the nativism of his master Müller, who considered most animal functions, including their sensory abilities, as innate and therefore potentially present since birth. Attracted by Locke's ideas, Helmholtz considered behavior and sensory activity far from innate, rather being acquired through experience and apprenticeship, which explained the importance he attached in his experiments to an empirical approach to the phenomena. Du Bois-Reymond, on the contrary, remained faithful to his master's nativist ideas and showed little favor for empirical theories in the interpretation of his own experimental results. We might recall that for the philosopher empiricism was a doctrine according to which experience was derived from experience, and thus in opposition to classic inneist rationalism. For the physiologist, empiricism opposed nativism or inneism not in philosophical terms, which were concerned only with knowledge, but in terms of natural science that dealt with biological mechanisms and their either innate or acquired causes. This distinction helps us understand how du Bois-Reymond could be paradoxically both empiricist philosophically and nativist physiologically.

There were equally interesting differences of opinion between the two friends concerning religion. Both were materialists and opposed to any form of metaphysical vitalism, but did God exist for them? Du Bois-Reymond was a self-proclaimed atheist but more through intimate conviction than logical necessity. Indeed if for him science might explain everything in terms of physics and chemistry, it would nevertheless never be capable of explaining various questions on a universal plane. We would never know the reply to seven enigmas that he described as follows: the nature of matter and force, the origin of movement, the origin of the simplest perceptions, the problem of freedom, the origin of life, the apparently appropriate organization of nature, reasoning, and language that stemmed from it.[9] His atheism clearly came up against the limits of what he considered to be insolvable by science, and it rather resembled agnosticism, that is, the impossibility to either prove or deny that Friedrich Engels[10] described later as "shame-faced materialism."

Contrary to his friend, Helmholtz never adopted a clear stance as to whether these famous enigmas would one day be solved. He preferred to state that all scientific research would have been in vain for him if he had not had the conviction of being able to explain everything ultimately. He worked "as if" almighty science was capable of resolving everything, and in so doing he subtly reinforced the credibility of his physical model of the world. Otherwise, we know little about his religious convictions because he attended church very irregularly and always remained allergic to theology. He never claimed to be an atheist but only a materialist out of respect for the scientific method. For his friend Emil he would have been close to holding pantheistic ideas.

Conservation of Force

If in retrospect du Bois-Reymond seems to have been a little less visionary and subtle than Helmholtz, he was without doubt a great scientist, faithful to his friends, and with great foresight in his collaborations. Hermann had written his treatise *On the Conservation of Force* in a few weeks, having often discussed the contents with Emil. His presentation at the Academy of Physics, suggested by Emil, had first of all astonished and then impressed his audience but for very different reasons. Some were full of admiration for the new perspectives that were opened. In contrast, some older members, such as Magnus and Johann Christian Poggendorff, feared that in the hands of the young physiologist mathematics and physics might once again be forced apart, as in the past, whereas it was more important than ever to reunite these two disciplines in a single scientific approach.[11]

Publication of the treatise in the *Poggendorff'schen Annalen* was refused for reasons of circumstance, and Magnus advised its author to publish it at his own expense with an editor of his own choosing. Du Bois-Reymond, who had been his intermediary with the senior members, was shocked by their decision but nevertheless advised his friend to follow their advice, insisting that he also publish the introduction to his text, which he quite rightly considered to be the most important part—a rather daring piece that Helmholtz the purist found too tainted with philosophy.

The treatise first of all established a clear distinction between theoretical and experimental physics. For the historian Fabio Bevilacqua,[12] who made a thorough analysis of it, the methodology that Helmholtz adopted to demonstrate his theory of the conservation of force depended on four interactive hierarchical levels. The first level was a hypothesis that excluded all possibility of perpetual motion and reliance on a Newtonian concept of force. The second was the establishment of the principle of conservation of force. The third was a description of various empirical physical laws, and the fourth was the study of natural phenomena such as electricity and magnetism.

The progression from the second to the third level emphasized the task of theoretical physics in deducing old or new empirical laws from the principle of the conservation of force. Experimental physics was illustrated by the progression by induction from level four to level three, from natural phenomena to empirical laws. For Bevilacqua, it was clear that Helmholtz considered that experimental physics ensured the intelligibility of nature, but that it was theoretical physics which would make it comprehensible, when the empirical laws fitted completely the principles it had established.

Recourse to Kant

The treatise *On the Conservation of Force* was preceded by an introduction on the transformation of natural forces, which Helmholtz presented as a purely physical essay but which was in reality a metaphysical theorum of the philosophy of science and a brilliant review of his own convictions concerning natural science. He began with two basic principles: first, "that it could not be possible by the effects of any form of combination of natural bodies to obtain unlimited work"; and second, "that all natural activity could be reduced to attractive and repulsive forces of which the intensity depends only on the distance apart of points that act upon each other."[13] So Helmholtz stated from the beginning that perpetual motion was impossible and that the forces he cited in his treatise were those of Newton.

There then followed a distinction between experimental and theoretical physics. The former had to "discover the laws by which the particular phenomena of nature are bound by general rules ... laws such as those of refraction and reflection of light." Theoretical physics "on the other hand seeks to study unknown phenomena through their visible actions; it seeks to define them according to the law of causality. These studies are imposed on us and justified by the axiom that all change in nature must have an adequate cause ... The final aim of theoretical science is thus to discover the ultimate unchangeable causes of natural phenomena."[14] The law of causality, cited by Helmholtz in relation to theoretical physics, was that of Kant, for whom the concept of cause was a metaphysical prerequisite to ensure the legitimacy and comprehension of nature. On the contrary, the causes invoked to explain the modifications of natural events observed in experimental physics were different because they were empirical and permitted the development of scientific activity.[15]

After this first reference to Kant, Helmholtz set about defining matter and force, which made him call on Kantian metaphysics a second time: Knowledge only warranted the name of true science if it dealt with the whole of its subject according to principles defined a priori.[16] First of all, he attributed a metaphysical status to matter and force because neither matter nor force were the direct result of experience, and one could only be aware of them indirectly by inference based on observation. Matter and force were therefore truly a priori transcendentals opening the doors of a possible understanding of nature. For Helmholtz, all matter was characterized by "its spatial distribution and its quantity (mass) that is posited as eternally unalterable." The only change that matter could undergo was its position in space, that is, motion, of which the cause was the ability to act or force. Matter and force were inseparable because they were two attributes of the same reality, two abstractions formed by the same intellectual process for one could only know active matter. In this context, Heidelberger noted that Helmholtz adopted a position of "realist metaphysics" because although giving the matter-force pair a metaphysical status, he was convinced that it was not fiction but that forces really existed behind matter. In conclusion, to understand nature meant to reduce natural phenomena to unalterable forces of attraction and repulsion whose intensity was dependent on the distance apart.[17] He added that the task of theoretical science would be fulfilled "if, at some time, the reduction of all phenomena to elementary forces were completed, and if, at the same time, this reduction were shown to be the only possible one that the phenomena permit. The necessary conceptual form for understanding nature would then be provided and objective truth would have to be ascribed to it."[18]

Fragonard's Swing

Who is not familiar with the light-hearted painting by that gracious artist Jean-Honoré Fragonard (figure 5.1), in which a young girl in a pink satin dress and silvery silk stockings is swinging elegantly beneath the foliage, glancing mischievously at a young man appearing from the bushes and spying on her? When we see a painting as joyous and carefree and so typical of the epicurean eighteenth century, it would seem incongruous to reduce the very instant that the two persons exchange looks to a simple law of physics. Yet we are tempted to imagine this instant, which is fixed forever in time, as reflecting the laws of the movement of the

Figure 5.1
Jean-Honoré Fragonard's painting of the swing (*Les heureux hasards de l'escarpolette*),
circa 1767. The young lady has been accelerating on her downward ride and is about to
reach her apogee where her kinetic energy will be nil and her potential energy at its
maximum before swinging back.
(Alinari/Art Resource, NY)

swing. Swinging in one direction and then the other, stopping at the top of its upward motion before setting out down again, accelerating to a maximum speed, then decelerating until it reaches a certain height and stopping, before starting another cycle. Even someone who is not conversant with physics can easily understand that after a moment of immobility at the top of the swing, the girl descends under the effect of gravity and picks up speed that allows her to climb up the other side, as if this speed enables her to overcome gravity. But it soon takes the upper hand and slows and stops her in full swing. So there are two forces in play: one related to gravity and the other to speed. The former is at a maximum when the swing is at its highest point, the second when it passes through the vertical at top speed. Everything seems to the naive observer as if these two forces have some mysterious relationship by which increase of one results in decrease of the other and vice versa. We are at the very heart of Helmholtz's problem.

He would certainly have pointed out that use of the concept of energy was preferable to that of force: Even if the latter was really the ultimate cause of motion, energy measured the capacity of a system to produce mechanical, calorific, chemical, or electrical work under the impulse of a force. This permitted the concept of the moving swing containing energy capable of producing mechanical work. Furthermore, according to our physiologist-physicist, this energy could be considered as the sum of two sorts of energy: kinetic energy ("vital force"—*lebendige Kraft*) and potential energy ("force of tension"—*Spannkraft*). The former was what the swing contained due to its speed, and the second is what it contained on account of its position in space where it was subjected to gravity, which accounted for its weight. Potential energy was maximal when the swing was at its highest vertical point and minimal at its lowest. The sum of kinetic and potential energy was always constant.

In Helmholtz's theoretical language, it was expressed as: "In all cases of movement of free material points under the influence of their attractive or repulsive forces of which the intensity depends only on their distance, the diminution of force of tension is always equal to the increase in vital force, and the increase of the former is always equal to the diminution of the latter. The sum of vital force and force of tension is always constant."[19] Helmholtz called this law the principle of conservation of force, where force really meant energy.

The naïve observer could object that Helmholtz's principle was inexact because if you left the swing to itself it would stop. Certainly, but he would have answered that the swing lost its energy progressively because it was slowed by the air that it disturbed and warmed and that the friction of the mechanical structures also caused loss of energy through heat which was lost irreversibly. The example of the swing is a fruitful illustration but demonstrates the difficulty in reconciling an empirically observed phenomenon with the abstract principle of the conservation of energy (the interaction between levels two and three mentioned earlier).

After Leibniz
In his manuscript of 1847, Helmholtz still used the terms "vital force" and "force of tension" instead of those of "kinetic energy" coined by William Thomson (Lord Kelvin)[20] and "potential energy" by William Rankine.[21,22] Helmholtz's terms were in the direct line of the tradition of Leibniz, who had used the expressions *vis viva* and *vis mortua* to mean approximately the same thing. Helmholtz's major contributions after Leibniz were to give to this pair the dimension of work, perhaps influenced by the definition of the latter by French engineers, and to quantify each term of the pair in compatibility with Newton's laws and his concepts of force.[23]

The consequences of his theory were fundamental. Leibniz had already invoked a principle of conservation to fit his conviction that one could neither create nor destroy work without compensation, but for him the *vis viva* could not be compared or equated with the *vis mortua* except qualitatively because they were two different concepts, one dynamic and the other static. When Helmholtz decided to quantify the two parts of Leibniz's pair in the context of Newton's mathematical and physical formalism, the way was prepared for the empirical establishment of a causal relationship in which variation in one would necessarily provoke variation in the other but in the opposite direction. A static cause (potential energy) could have the effect of generating movement (kinetic energy). This in turn might cause the return of the object to its initial position. From the moment the object left its initial position until its return, the sum of kinetic and potential energy remained constant, which fitted the principle of conservation of force, which could be inferred from the two propositions of the Introduction: the impossibility of perpetual motion and recourse to Newton's concepts.[24] Helmholtz had much merit that, in his enthusiasm, he proposed to use a single common measure for all natural phenomena in the shape of energy in both its static and dynamic forms. This was indeed a remarkable effort to find a unified theory to understand nature, but it was unfortunately utopian.

The Ambiguities of Perpetual Motion

The shrewd reader might ask whether, after all, a "perfect" swing in an enclosed space, moving in a vacuum and without friction between its mechanical parts, would be an ideal example of perpetual motion because in such conditions it would swing back to its starting point after each cycle in virtue of the principle of conservation of force. Certainly we might conceive of the possibility at least abstractly. There is another example, that of inertia, the property of bodies to maintain indefinitely the same speed imparted by a given force at a given moment. But here again, the motion is only perpetual in the abstract, in a closed system, beyond all other forces. These two examples made Pierre Costabel note that there was real ambiguity for classic science when it proclaimed the permanence of inertial movement while maintaining, in contrast, that perpetual motion was a figment of the imagination.[25]

We must recognize that in nature it is impossible to imagine a machine with perpetual motion, and the example of the swing eventually stopping due to friction and consequent generation of heat is there to remind us. It has always been so despite an inordinate amount of effort expended to achieve it. We may recall, for instance, the clever apparatus devised in the thirteenth century by Pierre de Maricourt, one of the first experimentalists of history, which was based on the bipolarity of the magnet.[26]

Furthermore, even in the utopian case of the "perfect" swing in an enclosed space, in a vacuum and without friction, it would be wrong to speak of perpetual motion because at the point between cycles it is immobile and only an external force can set it in motion again. Helmholtz would certainly have gone even further, judging from the terms he used in his own treatise: If the perfect, enclosed swing remote from external forces and always returning to the same highest point suddenly without apparent reason went a little higher, it could only be due to some theoretical force imparting supplementary energy to it. In such conditions, he would have said, there must have been creation of energy and work *ex nihilo*, and then one might talk of perpetual motion.

The paradox of admitting the theoretical possibility of a "perfect" swing while considering any possibility of perpetual motion as pure fantasy has its roots in philosophy. Aristotle already estimated that all movement that was not natural did not exist independently and only existed because of the exhaustion or consumption of something else.[27] But it was Leibniz who resolved this recurrent problem that dated from antiquity (is perpetual motion the expression of perfection?) by pronouncing a metaphysical principle according to which creation from nothing, *ex nihilo*, could not figure among the acceptable actions in a created world. The only valid, logical principle for phenomenological science was the total equation of cause and effect. This principle excluded any creation of work or energy from nothing because the effect would then be greater than the cause, which was impossible. The expression *perpetuum mobile* then found all its deepest meaning because even the "perfect swing" could not alter the height it reached without addition or subtraction of energy *ex nihilo* or *ad nihilum*. Leibniz's central principle was certainly crucial in Helmholtz's mind: He doubtless found in it the metaphysical basis needed to define the conservation of force and, in the process, to reject vitalism as a doctrine in which the vital agent in the fight against death was a force created *ex nihilo*.

The Essential Dialog between Theory and Empiricism

For Helmholtz, to understand nature meant reducing natural phenomena to inalterable forces of attraction and repulsion depending on the distance between the active elements. On the basis of his metaphysical prerequisites—the impossibility of perpetual motion and reliance on Newton's forces—he deduced the key principle of the conservation of force, which he confronted in the rest of his treatise with empirical data and rules derived from experimental physics. In doing so, he wanted to demonstrate that his principle, already used by Sadi Carnot[28] and Emile Clapeyron,[29] as he said himself, applied to all phenomena of which the laws were sufficiently certain, and he offered a beneficial guide for the experimental study of the others.

This program was more easily applicable to the world of mechanics, already thoroughly explored since Newton, than to that of light, heat, electricity, magnetism, or electromagnetism, of which almost nothing was known.[30] His strategy was to search for a mechanical equivalent for all these obscure phenomena and thus absorb them into his general principle of conservation of force. We can readily imagine, in the light of our modern understanding of physics, that it was not an easy task. However, thanks to his knowledge of natural science and his expertise in mathematics, he succeeded in unifying what could be unified in his mechanical model of nature and in indicating the experimental opportunities to follow up wherever current knowledge was sorely lacking. Helmholtz apologized to his readers for the large number of hypothetical elements, and he concluded that his "law was not in contradiction to any known scientific facts and that, on the contrary, it was confirmed by a great number of them."[31] Furthermore, he wanted to present to the physicist as completely as possible the theoretical, practical, and heuristic richness of this law, of which the definitive demonstration could be considered as a problem for the immediate future of science.

This physicist's *credo* did not enjoy immediate popularity except among his close circle, including his master Müller. We saw previously the negative reactions of Magnus and Poggendorff who feared a return to predominance by mathematics. To this was added, according to Helmholtz's biographer Koenigsberger, the fear that recourse to metaphysical prerequisites represented a final convulsion of Hegel's natural philosophy.[32] Numerous other criticisms were voiced, notably of plagiarism from authors who considered that they had said approximately the same thing. Despite the lively polemic that ensued, these accusations have been definitively refuted by historians. It is certain, and Helmholtz did not deny it, that many of the data he used were already known when he wrote his treatise. But his original contribution was his enormous effort of synthesis and unification of scientific concepts based on his reliance on Newton's concepts for all empirical aspects of natural science but subject to theoretical metaphysical prerequisites inspired by Kant's philosophy.

The more the years passed, the more physicists understood the meaning of Helmholtz's message; he, more than anyone else, had fought for the triumph of the principle of conservation of force as he understood it. Its acceptance by the scientific community was to bear fruit to an unimaginable extent for the natural sciences, including biology. As the physicist Max Planck emphasized fifty years later, one should not say that Helmholtz discovered the principle of conservation of energy, as if he had expressed the idea for the first time, because he, no less than his colleague Julius Robert von Mayer, who accused him of plagiarism, had a great number of predecessors in this domain. What was frankly new in his treatise was that it showed for the first time the significance of conservation of energy, almost unknown to physicists at the time, for the study of isolated physical phenomena and the quantitative consequences of its acceptance for different fields of physics. At that time, it was even more difficult to have a global view than was the case in Planck's day, but the concept had proved itself as knowledge had progressed.[33]

Goethe had tried to explain nature simply by the relationships between observed phenomena. On the contrary, by seeking an explanation in the play of invisible "atoms" activated by forces of attraction and repulsion, which by their interaction created confusion that was difficult to understand but nevertheless legitimate, Helmholtz distanced himself from Goethe. Indeed he did not hesitate to proclaim it openly a few years later at Königsberg in his famous speech devoted to the poet.[34]

The Manifesto of 1847

The treatise *On the Conservation of Force* appeared at the right moment, not only for the experimental project that Helmholtz had planned but also to support his conviction, and that of his friends, that physiology should henceforth steer a resolutely experimental and physicalist course. The primary aim of experiments was to find the causes of phenomena and not to test hypotheses, which was only of secondary importance. It should be an *ars inveniendi* and not an *ars demonstrandi*.[35] It must also be physicalist because only physics had established the major theoretical principles that permitted the quantitative unification of the phenomena of nature, including biology.

As co-actors in the reform of medical studies in Berlin (see chapter 4), Helmholtz, du Bois-Reymond, and Brücke wanted to ensure the development of an independent physiology as a sort of organic physics. In association with Carl Friedrich Wilhelm Ludwig,[36] a physiologist in Zurich and then in Leipzig, the quartet drew up the *Manifesto of 1847*, creating an organic physics exclusively based on the laws of physics, chemistry, and mathematics and in which physiology was reduced to its physicochemical fundamentals. The most spectacular outcomes of the manifesto were not only the adoption of a materialist path that henceforth the researchers of the institution would follow openly in the formulation of their scientific method, but also the almost explosive development of techniques and laboratory equipment conceived for the specific requirements of their research.

In the background to this courageous and enterprising group, and approving the direction they were taking, was Alexander von Humboldt, an attentive but mostly invisible spectator but always ready to intervene when one of his "Humboldt boys" needed a helping hand. In fact he would soon intervene effectively once again in favor of his protégés when an academic appointment at the Academy of Art became vacant.

Intermezzo with Artists

All arts begin with necessity.
—Johann Wolfgang Goethe

Emil du Bois-Reymond is credited with saying that conserving Helmholtz for science was as important as conserving force. This was far from
being a joke for meanwhile their friend Brücke had been offered the
chair of physiology in the University of Königsberg following the death
of Burdach. Brücke held the more remunerative than prestigious post of
professor of anatomy at the Academy of Art in Berlin, which had enabled
him to pursue his research in Müller's laboratory. So his post became
vacant and would normally have been filled by du Bois-Reymond, the
senior assistant. But he wanted to continue his research on animal electricity without wasting any time and, estimating that he was earning his
living well enough, proposed spontaneously to his younger colleague
Helmholtz to help him obtain a post that would guarantee his scientific
future.

Very happy with this agreement between his collaborators, Müller
wrote a long letter to the minister supporting the candidature of his
student whom he eulogized enthusiastically, praising his competence and
the multiple facets of his culture as a scientist whose research promised
important results. Helmholtz was one of the rarest talents he had known.
He recommended him as much for his studies of anatomy and physiology
as for his deep knowledge of physics. He wanted to seize any occasion
to permit Helmholtz to devote himself entirely to scientific research. He
recommended him wholeheartedly.[1] Events progressed rapidly: Müller's
recommendation of his student found favor with the minister, and the
candidate was invited to give a trial lecture at the Academy of Art. This
lecture, at the end of August 1849, was certainly a hard task for our
physiologist, who was more used to recording observations of biological

data in a physical context than an artistic one. However, he applied all the power of his logical mind, and it is rather poignant to note that in this lecture designed for nonscientists, he sketched the first timid outlines of his theory of perception. In addition, he described his admiration for the artists of antiquity striving for an ideal of truth that only genius could express.

He said in his lecture[2] that in Antiquity the genius of the artist was the mysterious power to discover and represent, intuitively and without calculation, that which a later critical examination recognized as true, complete, and justified. The emotion of the sensitive onlooker was all the greater the more richly and truly the artist had succeeded in perceiving and reproducing the idealized contents of his work. Failure in this task was perceived by the onlooker as an insult to the life and beauty of the whole work, even if he were incapable of saying where the artist had gone wrong and for what reasons. He remembered having seen in the Berlin Museum a statue of Apollo as an archer in which something indefinable prevented him from feeling unreserved admiration. He discovered that the imperfection was due to an anatomical error that he had not noticed immediately: The posterior part of the deltoid muscle of the shoulder was represented as if it took its origin from only a part of the spine of the scapula, whereas in reality it sprang from a wider origin. Thus, for the orator, the global perception of a work of art depended on unconscious visual analysis of details, the inaccuracy of one of which was capable of producing an unfavorable esthetic impression. This was the first time Helmholtz sketched the elements of his future theory of perception based on what he called "the unconscious inference of sensations," which was the true foundation for his psychophysiological work.

He went on to say that knowledge of anatomy never replaced genius or a talent for imitation a sense of beauty, but they could eliminate obstacles in the artist's path and give greater acuity to his inquisitive gaze. Every time an artist looked at a human body, he must "see through the skin" the degree of contraction of every muscle and the position of the relevant joints in order to give to his representation of the surface of the body the rounded, knotty, or smooth contours that corresponded to the anatomical truth, respecting above all an equilibrium that was close to ideal beauty. "The ostentation of the artist who has anatomical expertise in his figures, with which we can reproach even Michelangelo, and more especially several of his less gifted imitators, is just as unpleasant and far from reality as the lifeless, distorted forms produced by neglect of anatomical accuracy." This last statement seems astonishing to our present-day susceptibilities, but it shows to what extent Helmholtz's esthetic judgment was imbibed with classical, if not conventional, culture: He was certainly much more audacious in science than in art. This would be confirmed in his later work on the physiological theory of music.

Despite everything, his lecture was highly appreciated by the academic senate, and a few days later he received an offer of appointment to the faculty for the start of the academic year. At the same time, thanks to another intervention by his guardian angel Humboldt to the army minister, he was discharged from all military duties and demobilized, which permitted him to begin his academic activity with a totally peaceful mind while still remaining attached to Müller's laboratory for his research. However, this was not to last very long because a few months after his appointment, it was announced that his friend Brücke had accepted a chair at the University of Vienna, so vacating his chair at the University of Königsberg.

In Search of Lost Time

Berge ruhn, von Sternen überprächtigt;
aber auch in ihnen flimmert Zeit.

[The mountains rest beneath the splendor of the stars;
But even in them time ticks away.]
—Rainer Maria Rilke, "Wunderliches Wort"[1]

In search of lost time. … This is not Proust but the plan for Helmholtz's research program on neuromuscular physiology that he set himself when he arrived in Königsberg where he had just been appointed. The "lost time" was what he had been able to demonstrate between an electrical stimulus to a frog's nerve and the resulting muscular contraction. At the end of his research a year later, he was able to state to the Academy of Sciences in Paris, through the intermediary of his protector Alexander von Humboldt: "I found that for the nervous stimulation to arrive from the sciatic plexus to the gastrocnemius muscle of a frog a time delay is necessary that is not difficult to estimate."[2] So it was possible to measure the speed of propagation of the nerve impulse. This observation by Helmholtz ran contrary to the accepted ideas of the time and has remained one of the major discoveries of the nineteenth century.[3]

But what had happened since his appointment to the Academy of Art?

From the Banks of the Spree to the Quays of the Baltic

The musical chairs had continued. Helmholtz was scarcely established in Berlin in the academic post left vacant by his friend Brücke moving to Königsberg when the latter accepted an offer from the University of Vienna, so freeing his very sought-after chair of anatomy and physiology. Good strategist that he was, Müller again wrote to the Prussian minister of education and suggested *ex aequo* the candidatures of du Bois-Reymond, Helmholtz, and Ludwig. Ludwig was an excellent respiratory and circulatory physiologist and, what was more, was senior to his young colleagues. However, Helmholtz was chosen; despite his merit, Ludwig had committed the error of showing off too openly his liberal ideas at the time of the revolution of 1848. As to du Bois-Reymond, he preferred to stay in Berlin hoping, probably rightly, to fulfill his scientific ambitions in the wake of his mentor Müller. Helmholtz was appointed professor and could finally think of marrying Olga. The wedding took place in the Dahlem district of Berlin on May 29, 1849, after which the young couple left immediately for Königsberg, now the Russian city of Kalingrad. The town was situated in the extreme east of Prussia on the shores of the Baltic Sea. Its climate was appalling and its fog infamous. This had not prevented Kant from spending his entire life there, strolling every day until an advanced age without his immense intellectual creativity being affected in any way. But Olga suffered a lot from the atrocious weather, and after a few years her already fragile health worsened.

During this period, Hermann's father wrote numerous letters to his children. He asked Olga to encourage her dear Hermann to pay more attention to his tidiness because it was his weak point, and to suggest to him that when he became a father to be a better example of discipline than he had been himself. He recommended to Hermann never to forget the close ties between philosophy and physiology and to take advantage of his appointment in Königsberg to meet the famous philosophers who taught there, notably Johann Karl Friedrich Rosenkranz, Hegel's biographer and disciple.[4]

The university, called the Albertina, was founded in 1554 by Albert of Brandenburg, Duke of Prussia, for the deepening and propagation of the Lutheran faith. Its scientific eminence was considerable, especially since Kant, who had attracted many students, notably Herder. When Helmholtz arrived, he set about meeting his famous colleagues. In particular, he frequented the astronomer Friedrich Wilhelm Bessel, well known for his calculation of the trajectory of Halley's comet, who initiated him to observation of the heavens and made him aware of techniques to measure time and ways to avoid errors. He also had a close relationship with another contemporary scientist, Franz Ernst Neumann, a mineralogist and physicist, specialized in the optical properties of crystals and the electrodynamics of induced currents. His student was the already famous Gustav Robert Kirchhoff, well known for his work on electricity who later became his great friend. These close relationships formed at the

Albertina perhaps enable us to better understand the exceptional competence shown by the young professor of physiology in his subsequent study of the mechanisms of vision and audition.

Frogs and Muscular Work

It does not seem that Helmholtz already had the intention while still in Berlin to concern himself with the problem of the speed of conduction of the nerve impulse. As a result of his earlier experiments, it was rather the muscle that interested him as a paradigm of the transformation within a single organ of chemical energy to mechanical energy and heat. Furthermore, his treatise *On the Conservation of Force* had shown the importance of the notion of "work" for the measurement of energy, and he had concluded that the pursuit of his research on the energy of muscular contraction necessarily involved the measurement of the work done. Rather than using a continuous tetanic contraction of a muscle by electrical stimulation, which was an immobile and stable state, he preferred the brief and almost instantaneous contraction of a muscle raising a small weight attached to its extremity because only in these conditions could one speak of mechanical work, with the muscle moving a given mass a given distance in a given time.[5]

Such experiments were not too difficult to perform. The frog muscle was dissected with its motor nerve and suspended by one extremity on a hook, and on the other was attached a small weight. A brief electric shock was then applied to the nerve causing an almost instantaneous contraction of the muscle, which lifted the weight. To calculate the effective mechanical work, the physiologist had to quantify the time course of the contraction, which he tried to do according to a procedure developed by his friend Ludwig. He had measured fluctuations in intra-aortic blood pressure with a manometer, which transmitted variations in pressure to a membrane of which the reciprocating movements were coupled to a very light lever touching a cylindrical drum rotating at a known speed and of which the surface was "smoked" with a thin layer of carbon. The tip of the lever scratched this layer and marked a trace of time horizontally and amplitude of the pressure variations as recorded by the manometer vertically. This was the *kymograph* (figure 6.1).

Helmholtz used the same sort of apparatus but attached the end of the writing lever to where the weight was hooked to the muscle. In addition, he constructed a mechanism to enable the rotating cylinder to trigger the electrical stimulation of the nerve and record it on the smoked

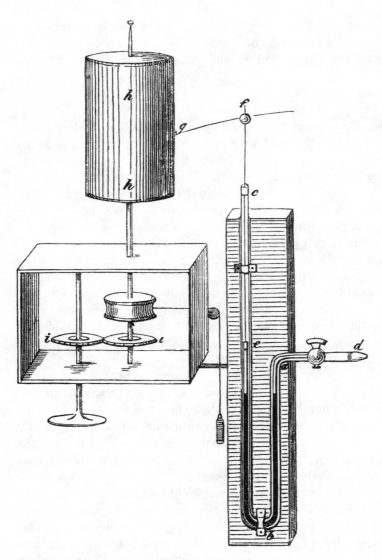

Figure 6.1
A kymograph, the predecessor of Helmholtz's myograph, as perfected by Carl Ludwig in
1846 for his studies of circulatory physiology.
(Ludwig 1858–1861)

surface. So the tracing on the cylinder had a horizontal component representing time, interrupted by a first vertical deflection marking the time of the electrical stimulation and then a second deflection slightly later describing the temporal course and the amplitude of the muscular contraction.

What he observed was to say the least unexpected: Far from being instantaneous, as appeared to be the case from purely visual observation, the muscle responded by a progressive rise in contraction followed, after reaching a maximum, by a slow decrease.[6] The observations were difficult, hardly recordable, and could not be quantified in any way. In particular, it was impossible to know precisely when the contraction began and so determine the time interval between the stimulation and the contraction. It was just at this time, when his research was still incomplete, that he and Olga moved to Königsberg.

Discovery of Conduction Velocity: Helmholtz's Double Strategy

An urgent duty awaited the young professor almost as soon as he arrived in the university city: to prepare his lectures. He set to work with great enthusiasm because he much enjoyed teaching and, as he said thirty years later,[7] it was in writing up his notes on nervous conduction that he recalled the often repeated statement by his mentor Müller that it was impossible with the available technical facilities to measure the speed of conduction of a nerve because it must certainly have been about that of the speed of light.[8] This excluded all hope of measuring it because one would have needed a nerve of enormous length, comparable to the distances needed to measure the speed of light, which was impossible to imagine. So he resolved to tackle this problem head on, and he began to develop an apparatus to measure nerve conduction speed that was totally different from what he had used in Berlin to study muscular contraction. By dint of much effort, he managed to obtain an experimental laboratory in the Albertina and set to work, assisted by his young wife who, surrounded by fragile apparatus and tangles of electrical wires, helped with the often delicate measurements with great enthusiasm. He even gave her the title of *Director of protocol for observed measurements.*

The principle of his new measuring system was based on an observation by the French physicist Claude Pouillet. He had noted that when a brief electrical impulse passed, the size of the deflection of the pointer of a galvanometer was proportional to the time of flow of the current. Helmholtz reasoned that he should be able to use an apparatus in which a current was sent to a galvanometer at the same time as the nerve was stimulated electrically and was interrupted at the beginning of the muscular contraction (figure 6.2). He would then calculate from the galvanometer reading the time from the stimulation of the nerve to the muscular contraction. To build this apparatus was far from a sinecure; in addition to his experience in physics and mathematics, he needed remarkable technical and manual skill to achieve it.

"In this way," he said, "I was able to determine by stimulating alternately the upper part of the nerve and the lower part that the contraction arrived a little later in the first case than in the second. The delay was visible by a greater reading on the galvanometer in the first case…. As the distance between the stimulated points on the nerve was 50 to 60 millimeters, the nerve impulse took between 0.0014 and 0.0020 seconds to travel this distance."[9] The speed of conduction was thus between 25 and 43 meters per second.

To confirm this remarkable result, Helmholtz had to resolve the problem of inevitable errors in measurement due to various causes, such as modification of conduction velocity by heat, the imprecision in the actual distance over which the nerve response was measured, or the weight lifted by the muscle. In this context, he was able to take advantage of his new relationships at Königsberg and, thanks to Neumann, apply calculations of probability that permitted increased reliability of the various measurements. For example, there was the method of the "least squares," already in use since 1810 by astronomers but rarely used by biologists, whose results were much more variable than in physics. Such techniques enabled the mean velocity of conduction in the nerve to be established at 26.4 meters per second.[10]

Through the intermediary of his friend du Bois-Reymond, Helmholtz asked Müller to present an urgent note to the Academy of Sciences of Berlin to establish the priority of his results. He made the same request via Alexander von Humboldt to the Academy of Sciences of Paris, of which he was a member. However, the discovery was so unexpected that Emil had to use all his diplomacy to convince Müller, whereas Humboldt resolutely refused to sponsor the communication in Paris. In fact the draft that Helmholtz sent to his friend was very badly formulated because it seems that only Emil managed to understand it at the time and realize the real sophistication of the work and the robustness of the physical and statistical methodology. As a true friend, he wrote a new version of the note, which this time Müller, and even Humboldt, not only understood but about which they were really enthusiastic. The former

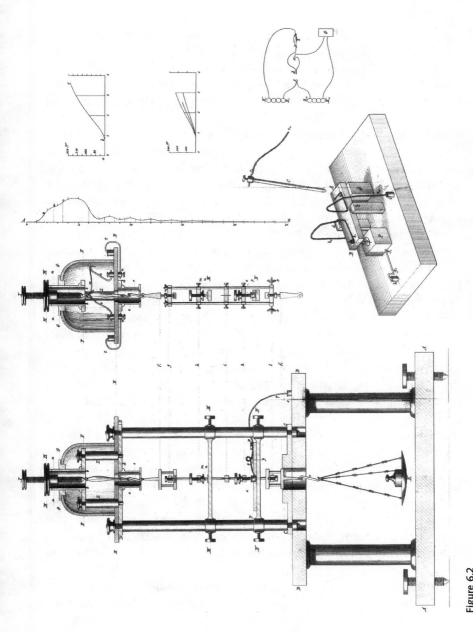

Figure 6.2
Apparatus derived from that of Pouillet used by Helmholtz for his first measurements of nerve conduction time. (Helmholtz 1850c)

suggested to his student that he extend his work by studying the reaction time of a sensorimotor reflex, which he later did. As to the second, he congratulated the young physiologist most warmly for having developed a great talent for experimentation using such delicate instruments, the great importance of this discovery being proved by the surprise it had provoked.[11]

Helmholtz learned some valuable lessons from these events. He realized not only that scientific writing was difficult and needed his whole attention, but also that the experimental method, and especially the mathematical analysis of data, had to be presented to the public in a comprehensible fashion if he wished to see his results reach more than a restricted circle of expert scientists and achieve wide recognition.

One is tempted here to recall what the French physiologist Etienne Jules Marey, a great admirer of Helmholtz, wrote about nerve conduction speed in his wonderfully pedagogical work *La machine animale*.[12] The text may seem somewhat naïve, but Helmholtz's concept was so novel: "Let us suppose that a letter is sent from Paris to Marseille and that we, as residents in the latter town, are informed of the precise moment when the postal train leaves Paris, while all we only know of its arrival in Marseille is the time of delivery. With this information how could we possibly measure the speed of the train? It is obvious that the time of delivery of the letter does not indicate the time of arrival of the train, for between its arrival and its delivery there are formalities, such as sorting and transport, which take a certain time that we do not know about. To have a more exact idea of the speed of the postal train we need to know the moment the train passes intermediate stations between Paris and Marseille, for example Dijon. Then we see that the delivery of letters is six hours later after passing Dijon than departing from Paris. Knowing the distance between these two stations we can calculate, from the time taken to cover that distance, the speed of the train. If we assume the speed to be uniform, we can calculate the time of arrival at Marseille, from which we can calculate the time needed for sorting and delivering the letters."

The ideal would have been to reproduce Helmholtz's results using a smoked drum, which would have had the advantage of the results being "visible" without needing to be subjected to complicated abstract calculations. However, one only has to recall the failure of his first attempts, while still in Berlin, when his graphic recordings were much too small to be useful. Despite that, he took up the concept of Ludwig's cylinder again and perfected it by making the rotational speed more regular and ensuring that the intermediate lever systems were as light and free from friction as possible. The *myograph* was born (figures 6.3 and 6.4).

So he was easily able to demonstrate to the public as well as to his students the "lost time" between a nerve stimulus and a muscular contraction. He was equally able to show that this time was longer the further the stimulating electrode was from the muscle and thus calculate the conduction velocity to a close approximation, although less accurately than with the galvanometer method. It was clear from his experiments that the "lost time" was only partly explicable by the conduction time of the nerve because there was a considerable time lapse between the moment when the nerve impulse arrived at the muscle and when the contraction started. This time was about 0.01 seconds,[13] and Helmholtz explained it by chemical processes preceding the contraction.

As to nervous conduction, its remarkable and unexpected slowness obviously excluded any explanation in terms of simple transmission of electrical current and even less so any evocation of the old "animal spirits." The mechanisms underlying the propagation of the nerve impulse were of course unknown. But for Helmholtz, there was little doubt that the impulse originated, as indeed du Bois-Reymond had suggested, as a result of a molecular rearrangement and that its propagation could be explained by a physical process, such as the propagation of sound in air or any other elastic matter or that of the chain of combustion in a explosive substance inside a tube.

The discovery that the velocity of the nerve impulse was measurable provided a solution to an important scientific problem but obviously created others. Fortunately, Helmholtz said during one of his lectures, the distance of the brain from the sensory periphery was not too long because the slowness of the nerve impulse, ten times slower than sound, would otherwise mean that sensations reached consciousness long after the stimulation of the sense organ. It was indeed astonishing to realize that, in a whale, injury to the tail needed more than a second to alert the brain and that it was just as long before the muscles could react to defend it.[14]

What for Helmholtz was simply one more biological enigma was, however, for his philosopher father a veritable traumatism. He wrote to his son: "Concerning your experiments, the results seemed to me from the outset a little strange, for I do not feel my thoughts and my bodily reactions as taking place successively but rather simultaneously,

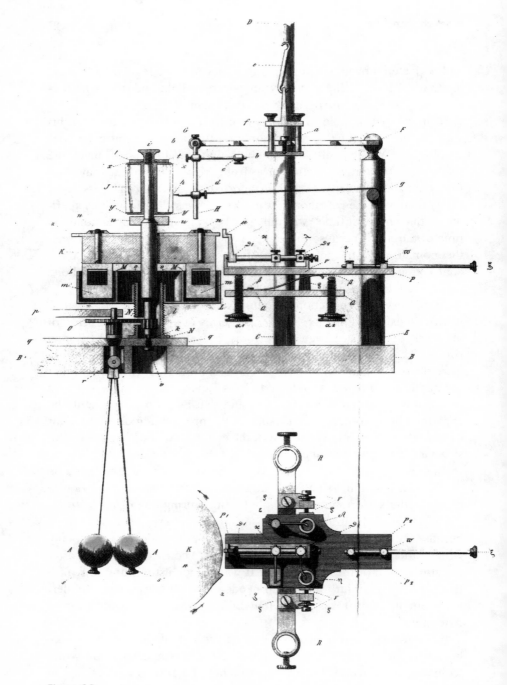

Figure 6.3
The myograph used by Helmholtz in 1852 to record nerve conduction time. He measured the contraction of the thigh muscle of the frog to electrical stimulation of its nerve at two different distances from the muscle.
(Helmholtz 1852a)

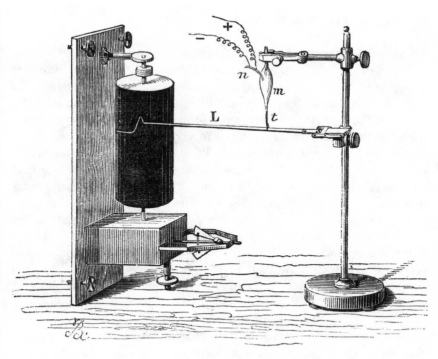

Figure 6.4
Recording of the contraction of a frog neuromuscular preparation obtained with a myo-
graph developed from that used by Helmholtz in his studies of nerve conduction time.
(Marey 1873)

as unities of the act of living, which after reflection become corporal and
spiritual: I would find it difficult to accept that a star that disappeared in
Abraham's time might still be visible today." His son replied very kindly
that he had no doubt "that the interactions between spiritual and corporal acts always originate in the brain and that consciousness and spiritual
activity have nothing to do with the time taken for information to come
from the skin, the retina or the ear, and that for the soul this nervous
conduction inside the body is just as good as outside like the conduction
of a sound from its place of origin to the ear."[15]

A few months later in 1850, in another letter to his parents, he had the
great pleasure to announce that he was the father of a fine and beautiful
daughter, Katharina Caroline Julie Betty, and that his wife, who had
helped him in the laboratory until the end of her pregnancy, was in excellent health.

Later Variations in Conduction Velocity

The measurement of nerve conduction velocity was not to remain without further work, as Müller had recommended, despite Helmholtz suddenly changing the thrust of his research as soon as the project was over. His idea was to concentrate on visual and auditory perception, and it was only much later, in 1867, when he was professor of physiology in Heidelberg that he returned to the subject, forsaking the frog for man.

At that time, in collaboration with Nikolaj Ignat'evic Bakst (Baxt), he observed that the delay between stimulation of a sensory nerve in man and the reaction in the form of a voluntary movement was on the order of a tenth of a second.[16] This time was longer than the sum of the successive conduction times in the sensory and motor nerves, even adding the "lost time" between the arrival of the impulse at the end of the motor nerve and the muscular contraction. This excess time was attributed to activity in the brain between the arrival of the sensory information and the departure of the motor command. The concept of "reaction time" was born, and one can foresee the later advances it permitted in neurology and psychology as a psychometric tool. Helmholtz had noticed very quickly in fact that this reaction time could vary to certain factors: It was shortest when the subject was "expecting" the sensory stimulus, but it was longer if the subject had to "decide" if a given sensory signal should trigger the motor reaction.

A few years later,[17] he wanted to know how long was necessary for a subject to consciously recognize a visually perceived object. To resolve this question, he used a so-called *tachistoscope* ("to see quickly") that he built himself (figure 6.5), adapted from a model by Sigmund Exner in 1868.[18] It enabled a subject to see in a flash (1/10,000 second) an image of some letters of the alphabet that he then had to identify verbally. The problem for the experimenter was that he wished to measure the time between stimulus and perception, the verbal description being merely a later proof of the actual perception. So how did he proceed?

The almost instantaneous visual stimulus was bright, and the subject's retina was dazzled and remained with a persistent image for a few seconds, during which the subject continued to see the stimulus even if he closed his eyes, rather as we experience the image of the sun after a brief glance toward it. This is the phenomenon of the "afterimage." So Helmholtz decided to vary the duration of the afterimage by "extinguishing" it with a very bright continuous light given a fraction of a second after the stimulus. The minimum duration of an afterimage in order for

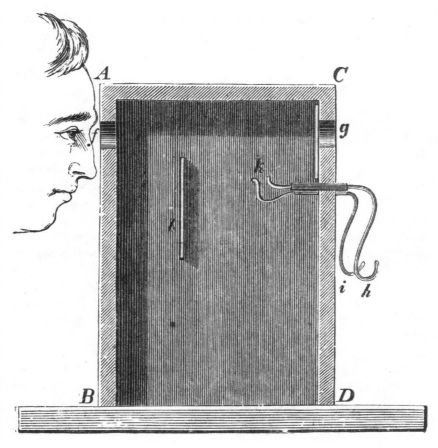

Figure 6.5
The tachistoscope, one of the instruments used by Helmholtz in his experiments on physiological optics. To avoid artifacts due to involuntary movement of the eyes, the test image was placed in a camera obscura and illuminated for a very brief time by an electric lamp. (Helmholtz 1866 III)

the subject to recognize the tachistoscopic image represented the time necessary to perceive and recognize an object. This experiment produced a value of about a thirtieth of a second. Remarkably, this time was far from constant and depended closely on the nature of the object to be identified, its complexity, and the subject's attentiveness.

This project was one of the last Helmholtz carried out on the physiology of the nervous system; afterward, he devoted most of his scientific research to physics, the old passion of his younger days. But to return to Königsberg in 1850: An important change of direction was being plotted within the solid walls of the Albertina. Our physiologist showed a sudden and unexpected interest in sensory organs, especially the eye. Then the drama unfolded.

The Invention of the Ophthalmoscope

Shortly before Christmas in 1850, he wrote to his father: "I have on the occasion of my lectures on physiology of the sensory organs made an invention which could possibly be of considerable use to ophthalmology. Actually, this invention was so obvious that it did not need any more knowledge of optics than I had learned at the *Gymnasium*. It now seems ridiculous to me that others and I myself could be so obtuse not to have found it earlier. It is a combination of lenses by which it is possible to illuminate the dark fundus of the eye through the pupil without using a blinding light, and at the same time to see details of the retina much more precisely than we can see the external features of the eye, without magnification, for the transparent parts of the eye act like a loupe of 20X magnification. One can see the most delicate blood vessels, the branches of arteries and veins, the entrance of the optic nerve into the eye, etc. ... I presented my invention to the Physical Society of Berlin like Columbus' egg, to be handled with great care, and I have patented it in my name."[19]

There can be little doubt that this invention was of considerable importance, but why had no one done it before him? Everyone is familiar with the strange glow in the eye of a cat or an albino rabbit when a light source near the observer falls on it. In the eighteenth century, this glow was still considered as produced by the eye itself, and some people believed that it was more intense when the animal was excited. It was only around 1810 that it was noticed that the eyes never shone in complete darkness and that the glow could be explained by a precise reflection by the *tapetum*, a reflective structure at the fundus of the eye.[20]

Although man does not possess a tapetum to reflect light, as in the cat and rabbit, Helmholtz's friend Brücke, as well as the English physician William Cumming,[21] had observed almost simultaneously that one could nevertheless obtain a faint reflection from the human eye as long as the light rays entering the eye were as parallel as possible to the observer's line of sight. However, they had not succeeded in seeing the fundus of the eye. In the absence of sufficiently accurate lighting, the pupil remained hopelessly black and seemingly opaque. Both men abandoned their project in despair. When he learned of these unsuccessful attempts, Helmholtz deduced that the human retina was capable of reflecting at least part of the incident light, but that to see the retina one had to first know the precise path of the reflected light, which Brücke and Cumming had failed to do. By applying his knowledge of optics and geometry, he was soon convinced that the reflected light followed the same path as the incident. It was clear to him that the pupil was black because the observer's head was necessarily always in the way of the incident light that, after reflection in the eye being observed, should have reached his own retina, thus enabling him to see that of his subject.

This conclusion was disarmingly simple and led him to construct an apparatus in which the key element was a small semi-reflecting mirror placed obliquely between the eye of the observer and that of the subject so that a laterally placed light was reflected by the mirror toward the subject's eye (figure 6.6a). The reflected light from his retina took the same path but in the other direction, and it divided at the mirror into two beams. The first was deflected back to the light source, whereas the other shone through the semi-transparent mirror to the observer's eye. However, even in these conditions, with Helmholtz's eye as close as possible to that of his subject, he still could not see the retina clearly probably because he was hampered by his own efforts to accommodate so as to focus on it. He overcame this by placing a concave lens between his eye and the mirror, and what he saw was truly amazing. The retina was pink, and magnified twenty times, with numerous blood vessels and the white optic disk where the nerve fibers from the retina left the eye to form the optic nerve. All he had to do was mount the mirror and lens on a convenient handle and the new *Augenspiegel* ("eye mirror") was born, later renamed *ophthalmoscope* by Marsilly in 1852. According to its inventor, he had only taken eight days to conceive and build the instrument (figure 6.6b).

One wonders whether Helmholtz was so accustomed to the microscope that he tried to fixate the subject's pupil as if it were a microscope slide. It is possible, but if so, he was complicating the problem because for an emmetropic observer (neither myopic not hypermetropic) it would have been enough to simply not accommodate, that is to say, to fix on infinity, for the concave lens to be superfluous for a clear view of the retina. On the other hand, if the observer was not emmetropic, correction lenses would be indispensable.[22]

Helmholtz was probably not the first to see the retina. It is said that Charles Babbage as early as 1847[23] managed to see reflected light from the human eye faintly through a small hole in the center of a mirror used to shine light into the pupil. This observation was not published at the time because Babbage was only able to see the retina on rare occasions; he had not thought to add concave lenses, which might have enabled him

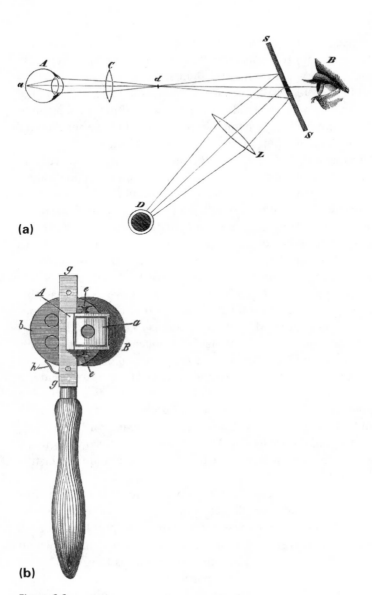

(a)

(b)

Figure 6.6
(a) Diagram of the observation of the fundus of the eye on which Helmholtz based the
construction of his ophthalmoscope. A, the patient's eye; B, the observer's eye; D, light
source; S, a semi-reflecting mirror (or one with a hole in its center) to allow the passage
of light rays from the patient's fundus to the observer's eye; C, L, convergent lenses;
C enlarged the image of the observed fundus, and L concentrated light from the source.
(b) Helmholtz's portable ophthalmoscope in its definitive form. The rotating disk had five
apertures: One was open and the others carried convergent lenses of different focal lengths
in order to focus on the patient's fundus and view it at different magnifications.
(Helmholtz 1856 I)

to focus the image of the fundus. At least that is according to Helmholtz, who only became aware of this observation several years later.[24] This emphasized yet again that an isolated observation, however potentially valuable, was of little use if not preceded by theoretical concepts that asked the right questions and if the experiments had no adequate physicomathematical basis accompanied by irreproachable technical methodology. Helmholtz possessed these qualities and so was rightly recognized universally as the father of the ophthalmoscope.

But there was more to be done to make his invention known. Once again Helmholtz left nothing to chance. The ophthalmoscope had been quickly recognized by the medical fraternity thanks to the proceedings of learned societies and scientific journals, but also by word of mouth between experts, which was, as today, one of the most effective tools for the dissemination of knowledge. The instrument had enjoyed almost immediate fame, just like Laënnec's stethoscope in 1816, which had made audible what was invisible. Now ophthalmologists had available a prodigious diagnostic tool, and the public, as well as political and academic decision makers, were aware of it. At that time, Helmholtz decided to make his first grand tour to the principal universities in Germany and Switzerland, his ophthalmoscope in his pocket.

At the end of the university year in late July 1851, he set off up hill and down dale, from Göttingen to Marburg, from Freiburg to Zurich and Vienna.[25] He met many colleagues, physicists, and physiologists, but also ophthalmologists. Everywhere he diligently demonstrated the efficacy of his instrument for the diagnosis of pathology of the eye that had been invisible until then, such as early signs of cataract, retinal vascular thrombosis, and actual neurological disease of the retina. He always met a warm welcome, so enthusiastic indeed that one day in a local newspaper, he was emphatically described as the emancipator and liberator of ophthalmology.[26] Many ophthalmologists asked him for his instrument, and before the end of the year, he had received 18 orders. This obviously enabled him to see the finances of his laboratory, which was in dire straights, in a better light.

He met some remarkable people, among them the physicist Wilhelm Weber, the specialist in mathematical optics Johann Bendikt Listing, and the strange philosopher Lotze. From Lotze, he later adopted the concept of "sign" in vision, as we discuss in chapter 8, and he described him as having much studied the principles of physiology and pathology, but too hypochondriacal and introverted to establish any form of conversation at an intellectual level.[27]

However, he did not neglect the cultural and touristic aspects of the countries he was visiting for the first time, and from then on he developed a love for mountains and long alpine treks. In his numerous letters to his wife Olga, who in the absence of her husband had moved back to her family home in Dahlem with little Katharina, age one, he alternated passages of exuberant affection with minute and enthusiastic descriptions of mountain landscapes such as the Rigi or the Gemmi Pass. During his journey, he stopped briefly in Strasbourg to see the cathedral of which he much admired the older "noble and imposing" parts conceived by Erwin von Steinbach. He climbed the tower where he could look out over the Rhine valley as had the young Goethe 80 years before him.

Helmholtz already enjoyed considerable renown in Germany and Europe thanks to the ophthalmoscope. Later he did not hesitate to admit that the invention and the associated media interest had contributed largely to ensuring the success of his scientific career, as well as the prestige and authority that he later acquired in German political and academic circles.[28] This was such that, in 1854, he was offered a Prussian peerage, which he declined because such responsibility required ambition that he did not have.[29]

The years that he spent at Königsberg from 1851 until his departure for the University of Bonn in 1858 were essentially devoted to research on vision. He published the results regularly in specialized scientific journals, but he also had the good sense to assemble them into a single voluminous *Handbook of Physiological Optics*,[30] which he re-edited and corrected several times throughout his lifetime and which became a work of reference for several generations of physicians. It contained just about the whole of contemporary knowledge concerning the eye and vision, from anatomy to the psychology of perception. The shadows of Kant and even Fichte were omnipresent because both philosophers were at the very roots of his own concept of perception. It was thus not surprising that he was invited to give the traditional lecture on the occasion of the celebrations organized by the university in Kant's memory in 1855, a lecture that he devoted to the relationships between the physiology of perception and philosophy.[31] However, his invitation on the occasion of the royal coronation in 1853[32] to speak on the scientific works of Goethe was more surprising because everyone knew of his strong criticism of the aberrations of the great poet in the field of natural science.

Before considering his contentions with Goethe's ghost, it is only fair to bring the latter back to life for a short while and allow him to speak because his thoughts on science were different.

7   Goethe and His Vision of Nature

Die Natur
Natur! Wir sind von ihr umgeben und umschlungen—unvermögend, aus ihr
herauszutreten, und unvermögend, tiefer in sie hineinzukommen. ...
Es ist ein ewiges Leben, Werden und Bewegen in ihr, ...
Sie verwandelt sich ewig, ...
Sie hat mich hereingestellt, sie wird mich auch herausführen. Ich vertraue mich
ihr. Sie mag mit mir schalten. Sie wird ihr Werk nicht hassen. Ich sprach nicht
von ihr. Nein, was wahr ist, und was falsch ist, alles hat sie gesprochen. Alles ist
ihre Schuld, alles ist ihr Verdienst.

[Nature
Nature! We are surrounded and embraced by her—unable to escape from her
and unable to enter more deeply into her ...
In her are eternal life, evolution and movement ...
She changes eternally ...
She has put me here, she will also take me away. I entrust myself to her. She can
rule over me. She does not despise her work. I did not speak of her. No, what is
true and what is false, she has said everything. Everything is her fault, everything
her merit.]
—Johann Wolfgang Goethe, "Die Natur," 1783

The scientific Goethe was a disturbing figure from the outset because his
strangely capricious mind flitted like a butterfly over the diverse mysteries
of nature that he endlessly explored over and over again. To expand
his understanding of nature was always an imperative task for him, as if
he found in this the indispensable breath of poetic inspiration. In his
works, poetry and science were always harmonious traveling companions.
Indeed Wilhelm von Humboldt, brother of Alexander, philologist
and founder of Berlin University in 1809, noted in his account of Goethe's
journey to Italy that poetic creativity in all true poets was always at the
same time a conception of the world.[1] Science and poetry were inextricably
interlaced because Goethe the scientist wanted to bear witness to

the unity of substance in nature as he saw it, and Goethe the poet found therein the expression of the word of God. He maintained that conviction to the end of his life and confided to Eckermann that Spinoza had always been for him a substantial support and that in him he found himself, for God did not only express himself in man but also in the richness and power of nature.[2]

His interest in science began around 1775 when he was appointed to the court of Weimar. He was entrusted with several missions, such as the renovation of the old copper and silver mines at Ilmenau in Thuringia, as well as the upkeep of the forests and pastures of the duchy. He devoted himself zealously to these new tasks, which allowed him to become familiar with the world of minerals and plants and awakened his scientific curiosity. Later he turned to osteology and color vision, which he studied conjointly with geology and botany, without ever losing his principal objective from view: to demonstrate that nature was unified and living.

The Quarrel over Basalt: Goethe between Neptune and Vulcan

Geology, or "geognosy" coined by Abraham Gottlob Werner[3] and the romantics,[4] was undoubtedly Goethe's first scientific passion. He first realized that in Ilmenau, and it developed during his numerous excursions in the Harz, in Bohemia, in the hills around Weimar, and finally during his journey to Italy, from the crossing of the Alps to his discovery of the great volcanoes Etna and Vesuvius. After two years of travel and minute observation in the peninsula, he returned to Weimar with his bags full of pieces of rock and specimens of minerals (figure 7.1).

For him granite was the archetypal stone, the foundation and bedrock of the Earth. Like an unshakable pillar, it supported the weight of the sea and constituted the framework of the mountains. Even if it was not the center of the Earth, it nevertheless formed the crust as far as one could penetrate. His first essay on granite[5] was more like a hymn to nature than a scientific publication. Several years earlier, he wrote in a flurry of enthusiasm, perched on the top of a "bare mountain," that he had contemplated the immensity of nature shrouded in mist and had imagined the land and the mountains emerging from the fury of the sea and life being born from apparent chaos. His impressions were recorded, as in his poems, in the first person, alternating his role between subject and object as we mentioned in chapter 2. According to his own account, it was at that very moment that he had the intuition of the essential role

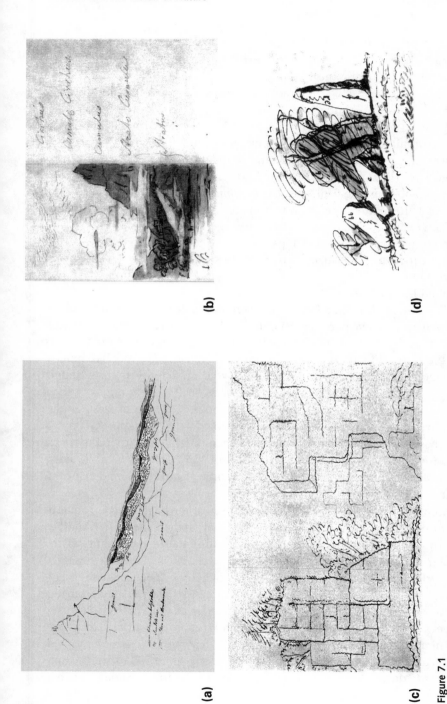

Figure 7.1
Geological sketches by Goethe (Femmel 1958–1979). (a) Section of geological stratification in a mountainous region near Karlsbad, 1806. (b) Similar mountain scene with clouds named: cirrus, cirrocumulus, cumulus, stratocumulus, and stratus. (c) Sketch of fractures in granitic rock, ca 1785. (d) A group of rocks at the foot of the Luisenburg near Wunsiedel where Goethe studied granitic formations and the effects of the atmosphere on the rocks, 1820.

played by granite in the formation of the Earth. One is tempted to compare this figure of Goethe surveying the scene with a painting thirty years later (1818) by the romantic Casper David Friedrich of a disheveled wanderer perched on a mountain top seeming both grandiose and yet tiny before the immensity of nature.

Later in his essay, he defined granite as a trinitary rock containing a uniform mixture of balanced proportions of mica, feldspar, and quartz. It was supposed to be derived from a very ancient process of crystallization when the primal seas cooled. As such it was remote from our senses and was therefore an *Urphänomen* or *primeval phenomenon*. After its formation, granite was transformed into various types of rock, such as clay, schist, or sandstone, and the scientist's role was to look for "transitions" from one type to another. Goethe sought to reveal these processes of transition by making beautiful drawings of broken and piled up rocks, trying to reconstruct in his imagination what they looked like before.

Soon after, in 1789, a famous quarrel took place over basalt, in which he played an active part the rest of his life. This quarrel had been brewing for many years and had its source in an old controversy about the origins of the Earth. In fact, 100 years before Goethe, after working like him on the modernization of mines and studying volcanic activity in Italy, Leibniz became interested in this question. He had concluded from his observations that the Earth arose from liquid igneous rock by a process similar to vitrification: The Earth's crust was a sort of vitrified matter. Glass was the basis of the Earth, and its debris formed the sand and then the rocks. Man lived on a volcano, and the center of the Earth was still liquid, a view close to that developed later by Georges-Louis Leclerc de Buffon[6] and Georges Cuvier.[7]

Leibniz's "vulcanian" theory was opposed at the beginning of romanticism by a very different theory, which soon gained the upper hand in Germany. It began with Werner, a forceful scientific personality at the watershed between Enlightenment and Romanticism, a deist obsessed by order and referred to as the Linnaeus of geology because of the rigidity of his analysis and classification of minerals. He rejected from the outset the idea of a fiery nucleus at the center of the Earth and refused to acknowledge the origin of rocks from volcanic eruptions. According to him, primal seas receded progressively, leaving fossils on the tops of hills and permitting the slow crystallization of granite and its transformation into different types of sedimentary rocks. As to volcanoes, they had arisen from the spontaneous combustion of underground coal derived from buried vegetation.[8]

Goethe acknowledged Werner's "neptunism" because the geologist's views agreed with his own observations at Ilmenau and in Bohemia, but above all because he preferred to believe in a progressive development of elements and forms. The process of their transformation for him was incompatible with volcanic catastrophes.

The quarrel about the origin of basaltic rocks was the opportunity for vulcanists and neptunists to debate their disagreements in public. Basalt had a volcanic origin for the former, but for the latter it was derived from the sea because it could be found at great distances from any volcano. Obviously Goethe opted for the neptunist theory but was nevertheless aware of the weakness of his arguments, especially as the young Alexander von Humboldt, after descending the Rhine to study the common basaltic rocks as well as the flora of the region, had openly supported the vulcanist view, taking with him many famous German geologists. Despite the gradual downfall of the neptunists, Goethe stuck to his opinions and suggested a compromise, which satisfied no one, in which basalt resulted from early sedimentation in the primal ocean brought to boiling point.[9] Until the end of his days, he remained a neptunist[10] and refused to admit the victory of his rivals. This demonstrated a sort of stubbornness that is strange because, in the end, Vulcan emerged victorious and opened wide the doors to modern geological science. Obviously Mephistopheles, who was a devil from Hell that he hoped to bring down upon Earth, was an advocate of vulcanism. Not wanting to accept this theory, Goethe made Faust emphasize the harmonious balance of nature because he did not believe in the explanatory force of the revolutionary eruption of magma.[11]

The Intermaxillary Bone or the Missing Link

It is quite moving to see how Goethe the poet failed in his attempt to become Goethe the geologist. His effort to unify natural processes was far from devoid of interest, but the rigidity of his romantic vision, and his aversion to violence and catastrophe in favor of intelligent and almost self-organizing chaos, prevented him from bringing the results of his strenuous labors and pertinent observations to full fruition. However, he enjoyed a remarkable revenge, admired by posterity if not by his contemporaries, in his research on comparative anatomy. He was introduced to this discipline at the same time as Herder during his stay in Strasbourg and had refined his knowledge at Weimar by reading and meeting the great anatomists of his time, and also in Padua where he attended sessions of dissection.

However, his research almost turned to disaster from the outset because he had been very impressed by his meeting with the theologian Johann Kaspar Laveter,[12] who initiated him to physiognomy, a study of correlations between character and the shape of the skull, which demanded a quasi-artistic gift of observation. Such an approach was likely to seduce Goethe the poet, who believed in an analogy between the shape of the skull and the great thoughts that it had enveloped during the lifetime of an individual. He had admired Raphael's skull, preserved in Rome, as a remarkable bony structure inside which a fine soul could wander with ease.[13]

Goethe the scientist nevertheless managed to tear his imagination away from such symbolism to turn to a problem of real importance, that is to say, the difficulty of reconciling the multiplicity of shapes in living beings with the unity of nature. He devoted the essential of his efforts to a comparative study of the mammalian skeleton. In his geological research, he was initially party to a theoretical idea about the formation of rocks from granite and then spent a large part of his life trying to support this theory. In contrast, in his anatomical work, he was more skillful and wrote an essay in a more academically correct style giving the results of his observations, then his interpretation, and then wrapping the whole into a theory.

But what had he observed? An apparently unimportant fact: All known mammals—for some he even cited Galen and Vesalius—had a distinct *intermaxillary* (or *premaxillary* in modern terminology) bone in the skull, wedged between the two halves of the maxilla, that carried the four upper incisor teeth (figure 7.2). In contrast, the existence of this bone was denied in man because the two half maxillas seemed fused together and bore the incisors like the other teeth. This official point of view was defended by eminent anatomists and was of great significance because it allowed the undisputable interpretation that man differed from monkeys in not possessing the intermaxillary bone that was present in all other mammals. Goethe's merit was to discover that this assertion was wrong and that man did indeed possess an intermaxillary bone because the sutures separating it from the neighboring bones remained visible in certain adult skulls, and especially because in the fetus the bone was easily visible and still quite separate. For Goethe, the presence of this bone demonstrated

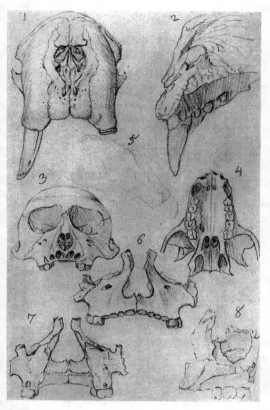 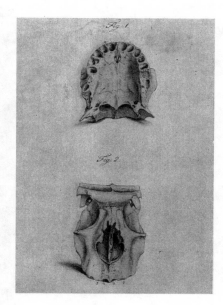

(a)   **(b)**

Figure 7.2
Figures from Goethe's study of the intermaxillary bone in vertebrates (Kuhn 1954–1964).
(a) The intermaxillary bone of a walrus, a monkey, and a man (not drawn by Goethe).
(b) A human intermaxillary bone seen from below and from above, respectively.

that one could not separate man from other mammals at this osteological
level. Man was not therefore "outside" nature, and the unity of nature
was perfectly real for those who knew how to recognize it.

However, his results[14] met with skepticism on the part of scientists,
many of whom were doubtless not yet ready to integrate man and nature
so easily. But he pursued his research and made a second very interesting
observation. During his journey to Italy, while resting on a beach near
Venice, he noticed the broken pieces of a sheep's skull scattered in the
sand. He had the immediate intuition, he wrote in his *Annals*,[15] that the
skull was formed by the transformation of vertebrae (figure 7.3).

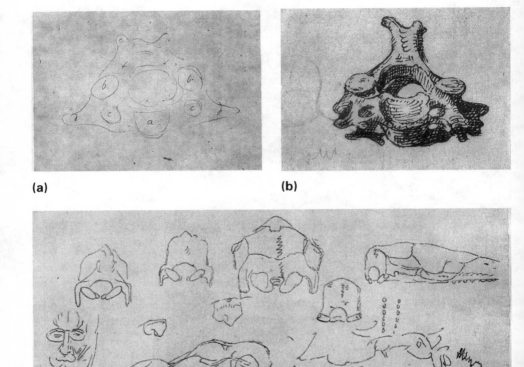

(a)   (b)

(c)

Figure 7.3
Drawings by Goethe from his osteological studies (Kuhn 1954–1964). Schematic (a) and completed ink and pencil drawing (b) of a human vertebra. (c) Skulls of various species viewed from different angles for his comparative anatomical studies.

Back in Germany and actively encouraged by the Humboldt brothers, he continued his research for a general organizational plan for mammalian osteology, in which man would no longer have a privileged position distinct from the animal world but would be solidly part of it. He developed the concept of the *Urbild*, or primeval form: All anatomical forms should be ultimately the expression of a generalized model, an abstract idea constructed in the mind. All mammals would undergo inevitable transformations from this original model, such as the emergence of the skull from vertebrae, and in this process man would not be better than or different from other animals.

At the end of his life, he still had enough energy to take the side of Geoffroy Saint-Hilaire, who, in a famous controversy with an empirical and inductive Cuvier, supported a synthetic and more deductive method thanks to which he penetrated deeply the spirit of the organization and forces of nature.[16] However, he did not experience the pleasure of knowing the eulogistic opinion that Charles Darwin expressed about him in the famous *Historical Sketch* at the beginning of the third edition of his *Origin of Species*, crediting him in 1794 and 1795, at the same time as his grandfather Erasmus Darwin and Geoffroy Saint-Hilaire and well before Lamarck, for coming "to the same conclusion on the origin of species." Darwin noted that Goethe recommended "that the future question for naturalists will be how, for instance, cattle got their horns, and not for what they are used."[17]

Sicilian Visions: The Primeval Plant

Goethe much admired Carl Linnaeus,[18] the great naturalist and botanist from Uppsala, but he consistently opposed his ideas. This love–hate relationship was understandable because he was never separated from his Swedish colleague's *Philosophia Botanica* of 1751, which he studied in detail and in which he had even found, to his great pleasure, some of his own theories. On the contrary, he estimated that Linnaeus, a man of the Enlightenment, had spent too much energy analyzing, naming, and classifying plants, which led to a static division of the plant kingdom. Goethe could not accept this in his desire to find, through observation of dynamic transformations of forms, convergent elements leading to a meaningful, unified theory.

He had always been interested in botany. While still in Frankfurt, he had collected a herbarium with his father's help and had expanded his knowledge at the University of Leipzig and in the field at Weimar. However, it was in Italy that he had his first intuitions about the nature of the process of metamorphosis that he developed into an essay[19] and several later publications dealing with biology (1817–1822). His *Metamorphosis of Plants* described in 123 paragraphs the growth of a plant from seed to fruit, as well as its own seed production (figure 7.4). From

Figure 7.4
Metamorphosis of a plant with its development from seed to leaf and flower: ink drawing by Goethe after 1790 on the basis of an inspiration during his visit to the Botanical Garden in Padua.
(Femmel 1958–1979)

the outset, he made the reader aware of the "regular course of nature" and "learn to know the laws of metamorphosis by which she … produces the most diverse forms by the modification of a single organ" (paragraph #3).

There then followed a description of the first organ to develop from the seed, the cotyledon, that he called the *seminal leaf* because it had the shape of a true leaf (#11 to #14). Goethe distinguished three successive periods in the process of growth in each of which there were alternate stages of expansion and contraction. "From the seed to the highest development of the stem leaf, we first noted an expansion. Then we saw the calyx develop by a contraction, the petals by an expansion, the sexual organs again by a contraction. We were soon aware of the greatest expansion in the fruit and the greatest concentration in the seed. In these six steps nature completes relentlessly the eternal work of the reproduction of plants through two sexes" (#73).

This metamorphosis in six stages, common to the whole plant kingdom, was however expressed very differently from one plant to another, and this presented Goethe with the same problem as in osteology—that of unity in diversity. During his journey to Italy, he was dazzled by the diversity of the plant world, and it was there that his botanical philosophy took its inspiration. He avidly visited the botanical gardens in every city, with his Linnaeus in his hand. In Padua, he was stopped in his tracks by the luxuriance and variety of the flowers, and he suddenly realized the importance of environmental factors such as climate, altitude, and proximity to the sea for the genesis of this diversity. His astonishment was even more during his stay in Rome, where he saw from his window superb clusters of pink flowers blooming before the leaves from the old stems of a Judas Tree. The shock was yet greater when he discovered the luxurious vegetation of Naples with its profusion of myrtles, pomegranates, lemons heavy with fruit, and silver-leaved olives. In Palermo, he again became obsessed with the fundamental unity of the plant world, and he began to dream of a sort of archetypal or primeval plant, an *Urpflanze*, from which all this varied foliage and spectacular floral abundance derived. The God of Spinoza, he said, was present in the "garden of the world."[20]

It was time to indulge in some philosophy, and Goethe did not deprive himself of it in his search for his *Urpflanze*. Despite himself, he gave up any thought of finding it in a botanical garden, so he defined it as a symbolic plant, an intellectual model, but he did in fact make an imaginary drawing of it. His friend Schiller, to whom he showed this sketch, insisted that it was in no way an experience but rather an idea, in Kant's sense, which could not be experienced empirically but which was necessary for the explanation of phenomena.

He was just about convinced by Schiller, but his imagination gained the upper hand. He had not given up his theory of a primeval plant because he "sought to explain all the organs that appear different … from a single organ, namely the leaf." Elsewhere he stated: "In … the leaf lies the true Proteus who can hide or reveal himself in all vegetal forms."

His lyricism even brought him to compare the successive expansions and contractions in the growing plant to the systole and diastole of the cardiac cycle and consider this alternation as a fundamental property of the living individual. Systole, through the contraction of the heart, pushed the circulation of the blood forward, whereas diastole, which followed, dilated the heart that therefore filled with blood. Life was therefore an alternation of contradictory phases: systole, conquest of the world and a thrust toward infinity, and diastole, concentration and withdrawal into oneself.[21]

Christiane Vulpius, who later became Goethe's wife, had the pleasure of being presented with a long elegy in 1798, repeating her lover's ideas on the metamorphosis of plants in the form of poetic metaphors, sometimes subtlety erotic, in which she easily recognized herself. For example: "All forms are similar and none is the same as another; so this chorus reflects a secret law, a sacred enigma."

Goethe was certainly as disturbing a scientist as he was original. Disappointed in geology because of a preconceived theoretical view that prevented him from making the most of his numerous observations, he made a breakthrough in osteology, thanks to his intuition, bringing man into the center of nature and its dynamic evolution. He was also a scrupulous and inspired researcher in his description of the metamorphosis of plants, passing from a generic description of nature to a modern genetic concept of organic life.[22] We owe to him, among other things, the first use of the term "morphology," which, in his own words, must contain information on the shape, formation, and transformation of an organic body.[23] This dynamic view of morphology that was still hesitant at the time of his osteological research really took shape in his botany when he introduced the concept of the primitive plant from which all others sprang thanks to the metamorphosis of single organ, the leaf. His opposition to a mechanistic vision of the world in which individuals remained stable again supported his concept of the unity of nature in continual transformation.

In a letter written in 1828, Goethe designated the two principal motors of nature: the characteristic *polarity* of romantic biology that, like a magnet, united two opposite extremes, such as life and death or systole and diastole; and *intensification*, the continual aspiration toward something higher, a sort of finalism at work inside matter.[24] However, we see in chapter 8, which deals with the quarrel over color, that for the progress of science, poetry and reality do not always make a happy partnership.

8   The Dispute about Colors: Goethe or Helmholtz?

Wär' nicht das Auge sonnenhaft,
Wie könnten wir das Licht erblicken?
Lebt' nicht in uns des Gottes eigne Kraft,
Wie könnt' uns Göttliches entzücken?

[If the eye were not of the sun,
How could we see the light?
If God's own force did not live in us,
How could the divine enchant us?]
—Plotinus, *Enneads,* cited by Goethe[1]

Goethe conducted his morphological research in relative peace despite the polite indifference or frank disapproval of certain of his contemporaries. But things were very different when he began to study optics and the perception of color. Wielding the oriflamme of one indivisible nature, he waged a veritable crusade against mathematics and laboratory apparatus. Only recognizing information acquired through the sense organs as trustworthy, he vented his wrath on the astronomer and physicist Newton, guilty in his eyes of a most damaging error for the human mind. He even compared him to an astronomer who on a whim wanted to put the moon in the center of our solar system and was obliged to make the Earth, the sun, and all the planets revolve around our satellite, hiding the error of his initial hypothesis by artificial calculations.[2]

What were the real facts?

Newton—The Origin of the Scandal

Was the Cambridge astronomer really conscious of bringing revolution to the world of science when he proceeded to reduce white light optically to the colors of the rainbow? Nothing is less certain. It is even unlikely that in his scientific endeavors he felt himself concerned by man's 30,000-year adventure with color. Yet intangible and transient at dusk, or inseparably embedded in a jewel or the petals of a flower, color has been of the greatest symbolic and even religious significance throughout history. The ochre of primitive tattoos or the cave paintings of Lascaux, the white of a church vestment, the black of mourning, the red that frightens, and the green of envy are all eloquent on its ageless role as a culturally significant vector and a factor of social cohesion. By cutting light like a common cake into as many colors as there are hues in the rainbow, Newton caused a scandal because henceforth the world of color was on the border between two cultures: physical science and human science.

Newton's doctrine was simple and based on experimental evidence.[3] He had noticed by chance that a prism, which he had just bought at a market for his research into refraction, split a beam of sunlight that fell on its oblique surface into a pattern of all the colors of the rainbow. It had long been known that a light beam passing from one medium to another (e.g., from air to glass) was deviated from its straight path by a process called refraction. But what was not known was that a prism was able to separate white light into rays that differed from each other both in their degree of refraction and their color: The most deviated rays were violet, the least deviated red, with the other colors of the rainbow in between. Like all good experimenters, Newton performed the necessary control experiment: Using other prisms, he succeeded in bringing together all the different divergent colored rays to form a single white light as at the beginning. He concluded that white light was heterogeneous, composed of multiple rays of different colors. A flower was red, he said, because its petals absorbed all the colors of the solar spectrum except red, which was reflected to our eyes and perceived as such.

The basic colors obtained through the prism were, for him, seven in number: red, orange, yellow, green, blue, indigo, and violet. It seems that Newton chose the number seven by analogy with the seven notes of the musical scale. Furthermore, he arranged the colors in a circle, coinciding their spacing with the intervals between musical notes. This attempt to find coincidences and analogies between light and sound was much in fashion at that time, and there even existed "light organs" based on such concepts.[4]

Goethe Fights Back: Indivisible Light

One is taken aback by Goethe's stubbornness in his study of color from his first trials of 1791 until the day of his death when he was still discussing it with his daughter-in-law, Ottilie. Was it the obsession of a poet traumatized by an analytical approach to light? Or was it the anguish of the sage before the instrumentalization of nature? All of that probably, and even more, for a profoundly human visionary who believed in a meaningful harmony ruling the relationships between man and nature.

Nothing illustrated this almost tragic aspect of his work better than his "confession" written in 1810,[5] in which he described his first encounter with Newton's prism. At the beginning of his research on color, a physicist friend at Jena lent him a prism so that he could see for himself the colors that Newton described. He absent-mindedly forgot about it and was most embarrassed when his friend, who needed it, asked him to return the prism as quickly as possible. He did not comply immediately and so was able to obtain an idea of its properties, but in his own way and rather hastily. Instead of respecting Newton's experimental conditions, he placed the prism close to his eye and looked at various objects against the white wall of his room. In these almost derisory conditions, he obviously did not discover the seven colors of the rainbow, but he was surprised to find that he could see bright colors around dark objects against the white background. Encouraged by his discovery, he concluded that the colors were not a result of decomposition of light but rather a dynamic tension at the border between the dark object and the white wall—the interface of the poles of dark and light.

The scene was set.

The Eye: Object and Subject

Goethe devoted more than 1,000 pages to color, but the essentials are to be found in his treatise on the *Theory of Colours* of 1810, of which the original German title, *Zur Farbenlehre*,[6] was subtler than its English translation because it suggested a concept of both theory and practical teaching. For Lacoste,[7] it was difficult not to detect religious or even mystical or occult overtones: "*Farbenlehre* designates rather ambiguously both a very ancient doctrine of color, a developing anti-Newtonian science, and a theory that must support artistic practice." Let us look more closely.

From the outset, we must consider as a major contribution Goethe's study at the beginning of his treatise of consecutive images and the phenomena of simultaneous contrast. He did not really discover them because they had been known for a long time, but they were considered as accidental or pathological, and he was the first to observe them systematically, predict their importance for painting, and give them a physiological meaning.

A consecutive image could be demonstrated by fixating a bright object such as a white disk for a few seconds. When he looked away from the object toward the dark side of a room, he saw a round image floating before his eyes, light and yellowish, of which the edges soon became purple (paragraph #40 of the *Theory of Colours*). This then progressively invaded the disk, after which the edge turned from purple to blue. Finally, the disk became gradually colorless starting at the periphery and then slowly faded and disappeared. The phenomenon could last several minutes because each time he closed his eyes and opened them again the disk reappeared. The consecutive image always followed the gaze, and if he moved away from the dark wall, the light disk grew larger, according to the laws of perspective (#37 and #22).

For Goethe, the phenomenon depended on the retina and was therefore physiological; it represented intrinsic activity of the eye in response to light. The eye was therefore object and subject, receiving the impact of the light image passively, but reviving it immediately it disappeared from the environment and creating a succession of colors that replaced each other. Using the Plotian expression of "solar eye," he emphasized the fundamental analogy that, in his opinion, united the eye and the light and permitted the microcosm to hope to reconstruct the macrocosm: Only like could know like.[8] It is clear that in this model, understanding nature is not reducible to the models of Galileo and Newton.

But the subtlety of Goethe's observations had many surprises in store, often of the greatest interest. One evening, while visiting a forge and fixating the red hot iron on the anvil, he was surprised to see that the consecutive image was green when he looked away at a lighted wall of the workshop but became red if he looked at a dark wall (#44). It was as if the red had "evoked" the green. On another occasion, at an inn he stared at a sprightly young servant girl who entered his room in the half-light. She had a dazzling white complexion and black hair, and she wore a scarlet bodice. As soon as she left, he saw on the white wall in front of him a black face surrounded by a light halo and a clear silhouette with bright sea green clothes (#52).

Systematic observation enabled him to state that black evoked white and red, orange, and violet evoked, respectively, green, blue and yellow, and vice versa. Goethe constructed a "color wheel" from these data that allowed complementary colors to be predicted.

From this color wheel, Goethe decided that the fundamental colors were red, yellow, and blue. Indeed his wheel illustrated that red evoked green, which was a mixture of yellow and blue; yellow evoked violet, made from red and blue; and blue evoked orange, made from yellow and red. For him the eye always saw the total picture and so completed the whole chromatic spectrum from the three fundamental colors (#50 and #60).

Simultaneous Contrast and Artists

If the eye was able to create complementary colors after seeing a light object, it was also possible as it fixated the object; but the evoked complementary color was produced around the image on the retina that had not been stimulated by the object. For example, when he fixated pieces of white paper against a yellow wall, he saw them with a violet tinge (#56): This was the phenomenon of simultaneous contrast. Goethe saw a courtyard paved with gray flagstones between which grass grew as intensely green when the barely perceptible evening light cast a reddish hue on the stones (#59). It was also possible to observe a similar phenomenon not simultaneously but consecutively: If he looked at a piece of bright orange paper against a white background, he could just see a blue color evoked over the rest of this background, but at the instant that the orange paper was removed, there appeared, as predictable, a blue consecutive image in the shape of the paper, and the white background became orange (#58).

These remarkable observations were taken up and extended twenty years later by the chemist Michel Eugène Chevreul,[9] director of the dye works at the Gobelins tapestry factory in Paris. He was unhappy with the quality of the dark colors of his carpets and blamed them on "modification that takes place in us when we perceive the simultaneous sensation of two or more colors."[10] He published a book that had considerable influence on the evolution of painting in the nineteenth century and even in modern times.[11] One may recall an important precept from it: "To put a color on a canvas is not only to paint everything that the brush has touched with this color, it is also to apply the complementary color to the surrounding space."

Goethe would certainly have been overjoyed to have known Chevreul's work and the support it provided for his own observations. It influenced painters from Delacroix, through Signac and Seurat, to Mondrian and Kandinsky. Vincent Van Gogh wrote to his brother Theo that he wanted "to express the love of two lovers by a marriage of two complementary colors, their mingling and their opposition, the mysterious vibrations of kindred tones."[12]

However, painters of the nineteenth century hardly cited Goethe as the source of their insight into contrast and color. They cited Chevreul and other theoreticians of art, such as Charles Blanc, Charles Henry, Ogden Rood, and even Helmholtz, as we see later. One of their excuses might be that Chevreul never cited Goethe, although he cited and discussed the observations of other older authors.[13] Georges Roque, a specialist on Chevreul, attributed this omission to the fact that he probably did not speak German, but also that he was selective even when citing French authors and that his strong personality incited him to make all his observations personally.[14] Paradoxically, it was at the beginning of the twentieth century, under the influence of the theosopher and theoretician of art Rudolf Steiner,[15] that Goethe's influence really began to be felt. In particular, Mondrian and Kandinsky found in Goethe, via Steiner, the idea of a spiritual approach through color.[16]

Colors and Polarity: Yellow and Blue

The study of complementary colors incited Goethe to acknowledge three fundamental colors—red, yellow, and blue—from which all others were derived. The phenomena that led to this discovery were, however, subjective because were the results of physiological activity of the eye. So what were the objective chromatic factors in the outside world that produced this ocular creativity? Here began the long polemic with Newton. From the outset, he considered it outrageous to want to analyze sunlight, before which Faust himself had looked away because it was so bright, as we saw in our *Prelude*. The only thing that counted was the interaction of light with darkness, and color was the objective expression of this interaction as long as the interaction lasted.

For Goethe the philosopher of nature, there was no doubt that the tandem light–dark was a polar structure and formed a primeval phenomenon that defied analysis. But how could this dual interaction that generated objective colors explain the fact that the eye created subjective colors on the basis of three fundamental colors? Faced with a dialog with Goethe the philosopher, Goethe the scientist capitulated. As usual, he made some superb observations in the field but interpreted them without convincing others and thereby unfortunately lost much of his credibility in scientific circles.

The point of departure of his research was obvious: The interaction between light and dark took place at some distance from the eye, and the colors that resulted only reached the eye after crossing a more or less translucent or, as Goethe called it, turbid (*trüb*) medium. The purest form of turbidity was transparence, and between transparence and milky-white opacity there were an infinite number of degrees of turbidity. The concept of turbid was very symbolic in Goethe's ideology. It also designated the obscurity that stemmed from a confused mind and disordered thought, or the look that veiled itself in sadness and melancholy, as at the moment of the poet's farewell to Friederike Brion.[17]

So, the radiant and colorless light of the sun seen through an even weakly turbid medium looked yellow, and if the turbidity increased, the light gradually took on a red tinge that might deepen to ruby red (#150). Such was the case of the sun becoming red when seen through a thick layer of mist. However, if one looked at relative darkness through a turbid medium that was nevertheless partially lit, there appeared a blue color that turned to violet when the turbidity changed toward transparence. Such a blue color was seen from the top of a sunlit mountain when "only a few light vapors float before the infinite darkness of space" (#151 and #155). Through a turbid, but lit, medium, light was seen as yellow and dark as blue: These were two fundamental colors discovered by Goethe the physicist, wholly in agreement with Goethe the philosopher of nature. The polar structure of light and dark could in fact only give rise to another polar structure made from two fundamental colors. These were yellow and blue, which when mixed permitted one to obtain all other colors. So yellow and blue had a precise significance within the polar structure (#696). Yellow was positive: activity, light, bright, force, heat, proximity, and repulsion. Blue was negative: deprivation, shade, dark, weakness, cold, distance, and attraction.

However, there remained a major difficulty already mentioned earlier: how to reconcile these two basic colors of the objective physical world with the three fundamental colors of the subjective world of the eye, red, yellow, and blue? How to introduce red into the physical world so that, without being fundamental, it nevertheless enjoyed a special status? Goethe needed to find a trick. As if pulling a rabbit out of a hat, he introduced a strange concept that he had often used in other research—"intensification." Red was born as a result of the separate intensification of yellow and blue and their union to form a new color, red, in which the specific properties of yellow and blue had been lost (#697 to #699).

Colors and State of Mind

Goethe recounted in French the story of a high-spirited Frenchman who one day "claimed that the tone of his conversations with Madame had changed since she changed the color of the furniture in her room from blue to crimson" (#762). He used this as an illustration of the fact that the colors around us influence our state of mind and morality. Yellow was close to light. When pure it was serene, gentle and playful; on satin it was dignified and magnificent. It always gave an impression of warmth and comfort. Blue could not escape its origin from shade and darkness; it seemed to draw us toward obscurity, and even when not unpleasant usually appeared cold and sad. As to red, still referred to as purple, it was noble, serious, and dignified; made from the intensified poles yellow and blue, it brought an ideal of satisfaction. Green, from a mixture of yellow and blue, brought a balance between opposites, and so rest and peace (#765 to #802).

Goethe the poet paid close attention to these physicopsychic effects and gave free rein to his intuitive vision in his description of the influence he attributed to color. This was doubtless his poetic imagination, far removed from science, but so subtle and elegant in the person of this great connoisseur of the affective potential of the mind. Even if the *Theory of Colours* was a failure in the eyes of scientists, it has remained an endless source of study and inspiration for our best contemporary artists.

Therefore, one should not be surprised, from this brief review of Goethe's scientific achievements, that everything was in place to provoke their rejection by scientists, even forty years later when his theories were still being discussed in cultured German circles.

Helmholtz Shows the Way

For Goethe, discussion about color was always a perilous enterprise, and in this context he liked to cite a predecessor: "If you show a red rag to a bull it becomes angry, but a philosopher begins to rage as soon as you merely speak of color."[18]

Helmholtz was ready to enter the fray in this quarrel about color that was as ancient as the world itself. In fact, he was attracted by the overall problem of visual perception, with its background of numerous controversies shaking scientists, poets, and philosophers. The problem of color vision was his first battle field because he was irritated by Goethe's color doctrine that, in his opinion, reeked horribly of metaphysical vitalism. Twenty years after the death of the poet in 1832, this doctrine was hardly taken seriously by true scientists. However, for Helmholtz, it still constituted a real danger because of the esteem for this irrational concept of nature by the theosophers, as well as several idealist philosophers in the wake of Hegel. Once chased out of the door, it must not reenter through the window to contaminate science yet again.

Helmholtz's Praise and Criticism of Goethe

It seems strange indeed that Helmholtz should have been invited in 1853 to speak about Goethe's scientific work on such a solemn occasion as the festivities for the anniversary of the royal coronation, although his critical attitude toward the great man of Weimar was known to all. Did someone wish to cause him trouble? On the contrary, was it to mark him as one of the university's most forward-looking scientists? No one knows, but what is certain is that Helmholtz made the most of the occasion to play the role of the great statesman, not sparing his criticism but very diplomatically enveloping it with an air of serene comprehension of the visionary poet.

From the beginning of his lecture,[19] he spoke with sustained praise for the man, his obstinacy, and his enthusiasm for scientific research. He credited him with two major scientific contributions. The first was to have demonstrated that anatomical differences between animals were often variations of a common pattern induced by specificity of behavior, habitat, or feeding. The discovery of the intermaxillary bone in man was very important in this regard because it demonstrated the kinship of the human skull, devoid of a muzzle, with that of other mammals. This represented a major impetus given by him to the concept of comparative anatomy that was to prove so fruitful later.

His second contribution was to emphasize the analogy between different parts of the same organism, along the lines he had described for homologous bones in different species. Plants had captured his attention with the infinite diversity of their leaves and flowers, and he had studied the successive metamorphoses from the seed to the next generation. Having said that, despite the beauty and attention to detail of his observations, the idea of common ancestral organisms in plants and animals was not very useful for the advancement of science, the orator stated.

After his initial eulogy, Helmholtz became more caustic, recalling Goethe's first trials with the prism, his enthusiasm when he saw bright colors along the border between dark and light surfaces, and his stubborn refusal to listen to his friends that these phenomena could be completely explained by Newton's theories. The orator then exploited a sentimental argument, claiming to be really sorry to see how so remarkable a man as Goethe could have claimed, without proposing any credible argument, that Newton's conclusions had been a mere tissue of absurdities. He added that such antagonism as a matter of principle led him to believe that Goethe's mind, made inflexible by his irrational convictions, worked differently from that of his contradictors.

What made Goethe different from the others? Above all, he was a poet and an artist. What was important in art was to transmit an idea in words, sounds, or colors. Further, this idea dominated everything but remained mysterious even for the artist himself, who had not conceived it rationally but through the intuition of an eye within himself and his alert affectivity. Such an idea was obviously the opposite of the abstract ideas conceived by a scientific mind searching for universality and intelligibility. So there was nothing surprising in Goethe's affinity with Schelling's natural philosophy where nature revealed itself directly to the intuition of the researcher. Also Hegel's fervor to defend tooth and nail the poet's scientific concepts was quite understandable.

The accuracy and truthfulness of Goethe's botanical descriptions led him to state that the flower was derived from the leaf. But a true scientist would only see in this "law" a representation or intuition of indefinable and even esoteric nature. The observations of the botanist poet, although subtle, did not allow the formulation of a simple, universally intelligible rule. It was shocking to see how Goethe discussed the mixture of fundamental colors, whereas Helmholtz was obliged to construct complex apparatus to allow him to isolate these colors correctly before trying mixtures. It is true that for Goethe the poet the use of laboratory apparatus meant torturing nature, whereas one should treat it with respect and consideration like a work of art. Nature only revealed its secrets if it had total freedom because it was the transparent representation of an ideal world that we could only appreciate through our senses.

On the contrary, Helmholtz continued, the world of the physicist was that of invisible atoms, propelled by forces of attraction and repulsion of which the numerous interactions, although governed by strict laws, could hardly be comprehended globally. Furthermore, for the physicist, sensory impressions did not constitute an absolute authority for knowledge of external reality. Was not the same sunbeam light for the eyes but heat for the skin? Were our sensations anything else but signs of objects outside ourselves, like written characters or articulated words that one used to name objects? In the end, Goethe's scientific work appeared like a desperate attempt to save sensory infallibility that was under attack by science.

He concluded with due pomp that the true philosopher of nature sought for the levers, cords, and pulleys that worked behind the scenes to change the decor. To see this machinery obviously spoiled the beauty of the decor, but the scientist still pushed forward. However, the poet sat in the theater and did not see the machinery and would be only too happy to deny its existence and believe that such apparatus could only be the fanciful product of a pedantic brain: The decor was capable of changing itself unless it was activated directly by the "ideas" that were behind the artistic work being played out.

Finally, he said that one could not triumph over the mechanisms of matter by ignoring them, but rather by submitting them to the scrutiny of our intelligence. One had to become familiar with levers and pulleys in order to submit them to our scrutiny even if that disturbed the poetic contemplation of nature. This was the price to be paid to realize the true significance of physical research for the development of civilization and the full extent of its justification.

Light Waves: Waves in the Ether

When Helmholtz undertook the study of color, he recalled from the start what the contemporary physical concepts about light were. For Newton, as for Descartes, light was corpuscular in nature, and it was the English physician Thomas Young[20] who discovered that one could resolve a number of problems in physics if, instead of this concept, one substituted a wave theory, originally suggested by Christiaan Huygens.[21] If light was a series of corpuscles, Young said, why were only some of them refracted by a lens and others reflected? He had much difficulty in getting his ideas accepted, and it was thanks to the young French engineer Augustin Fresnel,[22] author of a theory of interference of waves, that he finally succeeded in convincing his peers in 1816.[23] Thus, it was on this firm basis that Helmholtz commented on light as a wave phenomenon on the occasion of his inaugural lecture as full professor in 1852.[24]

Light was a vibratory motion that was propagated in the form of waves in a hypothetical medium, the ether. Contrary to sound waves that oscillated longitudinally (i.e., in the same direction as their propagation), light waves oscillated perpendicularly, like waves in the sea. Furthermore, light propagated through space at the great speed of 40,000 miles per second, whereas sound was limited to 1,058 feet per second. That was why we saw distant events before hearing them. The Prussian mile being some 7.45 kilometers, this figure corresponds to about 298,000 kilometers per second, close to what is considered to be correct today.

Another difference is the frequency of vibration of the waves. The light spectrum was very broad, between 450 million cycles per second for red and 790 for violet. As to sound waves, they had a frequency from 16 per second for the lowest note up to 16,000 for the highest. Curiously, here Helmholtz went back to similarities between sound and light that had attracted Newton and that still interested his contemporaries. In each octave between low C and high C, he said, the frequency of the note doubled, which gave 16 octaves in total for the sound spectrum. Now if one used this criterion—doubling the frequency each octave—for the light spectrum, one only obtained a little less than one octave. What was more, at the extremities of the spectrum red and violet occupied several tones, whereas in the middle of the spectrum all the many shades from yellow to blue only occupied half a tone. He concluded that, as a result, it was in the middle of the spectrum that the eye best perceived variations in frequency of light waves.[25]

This enumeration of analogies between sound and light was at the origin of much research on aesthetic invariants common to the two domains. For example, Friedrich Wilhelm Unger[26] proposed a theory of aesthetic harmony of color, in which the most perfect musical harmony, the major chord, corresponded to the famous color combination of the Italian painters: red, green, and violet. Helmholtz pointed out that, according to his own calculations, the Italian combination did not correspond to any major chord. In contrast, red, green, and blue were Young's three fundamental colors, and this was perhaps the reason that their combination was so harmonious. Furthermore, he suggested that this game of analogies between sound and light should be abandoned because it was too arbitrary.

But there was another difference, and an important one, between the respective worlds of sound and light.[27] When several musical notes were played at the same time, as in a perfect chord C,E,G,C, it still remained possible to hear each note distinctly and separately. On the contrary, a mixture of colors could not be analyzed by eye: It was seen as a single color whatever its composition, and one could not detect whether a white was made from blue and yellow or from red, green, and violet. It was notably this apparent simplicity of color perception that had precipitated Goethe's total rejection of Newton.

The Young–Helmholtz Theory of Color

Helmholtz respected Newton, but he did not hesitate to deride him in passing. He claimed[28] that Newton had been mistaken in believing an old rule that a mixture of two colored pigments gave the same final color as the mixture of two similar-colored lights. Such was often true, but not always. If such generalizations were still current, it was because, although his method was flawed, the authority of the great scientist was able to sustain the idea for so long. Indeed Helmholtz had perfected a very precise technique that enabled him to isolate each component color of the solar spectrum and then mix them as he wished in order to create composite colors. Thanks to this, he noted, for example, that he obtained green by mixing blue and yellow pigments but white if he mixed light beams of the same blue and yellow. This showed, he continued, that one had to distinguish between additive colors obtained by mixing two isolated lights of the spectrum, and subtractive colors, the product of a mixture of pigments. In the latter case, the pigment absorbed part of the light shining on it while the rest of the light was reflected to the eye without being absorbed. The pigment's perceived color was obtained by subtraction of the light absorbed by the pigment—in other words, the color that had not been absorbed.

So, Helmholtz studied systematically the mixture of pure colors using his skill in arranging prisms and mirrors. This work[29] was an extension of that undertaken by Mikhaïl Lomonossov in 1756 and George Palmer in 1777, and especially that of Young who in 1801 presented to the Royal Society a remarkable trichromatic theory of color.[30] For him there were three fundamental colors: red, green, and blue. By mixing these in appropriate proportions, all the others could be obtained. He was convinced that this property of generating other colors could not be explained by the physical properties of light but in the physiology of the eye. It was impossible to imagine, he said, that for each shade of color there was a corresponding specific "resonator," and it was much more probable that the eye possessed three types of receptors, each corresponding to one of the three fundamental colors. In itself, light was a colorless energy: It was the eye that enabled colors to be perceived thanks to its specific resonators. One must admire Young's foresight and its part in the famous theory of specific energy that Johannes Müller elaborated a few years later in 1838 and that we describe further in the next chapter.

Helmholtz much admired Young, whom he found to be ahead of his time. He accepted Young's conclusions but encountered enormous obstacles when he in turn tried to study composite colors. Indeed, he only found a single pair of colors that gave white on being mixed, and he had difficulty reproducing all the colors of the spectrum by mixing the three fundamental colors. These unexpected difficulties no doubt stemmed from his use of much more sophisticated experimental apparatus than that of his predecessors. The mathematician Hermann Günther

Grassmann came to his rescue in 1853 but not without sharply criticizing his physiologist colleague's experiments,[31] reminding him that each color was characterized by a wavelength, an intensity, and a given degree of saturation: A color was less saturated the more white it contained. He completed his comments with a mathematical theory that it should not be difficult to find many complementary colors that gave white on mixing as long as the previous parameters were taken into account. Helmholtz returned to his experiments[32] after modifying his technique and, by taking into account degrees of saturation, succeeded in finding several pairs of complementary colors. However, he encountered a new difficulty because he was unable to understand why apparently physically pure colors appeared unequally saturated. For example, yellow was much less saturated than violet. He was beginning to wonder how to achieve a more quantitative colorimetry that would be the key to the problem when he received the news that James Clerk Maxwell,[33] a young Scottish physicist, had solved the problem. Helmholtz probably thought at that time that his world was collapsing around him, but his strong personality enabled him to resist. So what had Maxwell found?

In parallel with his work on Saturn's rings that made him famous, Maxwell succeeded in building one of the first apparatus for quantitative colorimetry worthy of the name (figure 8.1). At that time, it was obviously inconceivable to measure light energy absolutely, but important data could be deduced from an observer comparing one light field with another, which served as a standard. Using a fast rotating disk bearing three colored segments at its periphery, he produced a composite color that could then be compared with a standard color, or white, placed in the center of the disk. By modifying the area of each color segment until their fusion when the disk rotated gave a white equal to that in the center, he was also able to measure fairly precisely the amount of white that unsaturated some colors more than others and reconcile in the end Young's red–green–blue triad.[34]

Fortunately, Maxwell's brilliant ideas went in the same direction as Helmholtz's. The latter was thus able to benefit from them and include them easily in his hypothesis that was henceforth ready. In his *Physiological Optics*, he orchestrated his efforts to make the most of others' findings, hiding any possible embarrassment in masterly fashion: He described the history of the science of color, recalling the contributions of other scientists and giving an exhaustive account of his own research, illustrated by impressive mathematical formulae. He described Young's work

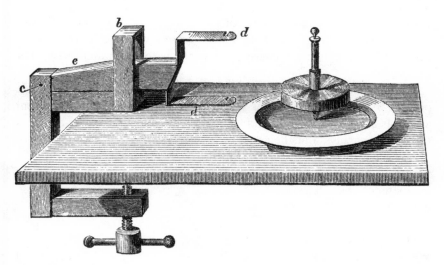

Figure 8.1
Apparatus with a rotating disk used by Helmholtz in his experiments on the fusion of
colors and the analysis of persistent visual images.
(Helmholtz 1860, II)

but in addition proposed his own personal synthesis, carefully defining
the limits between the respective fields of physics and psychology.

The reduction of light to three fundamental colors could only be of
subjective significance, he stated. The problem was to reduce three sensations of color to three fundamental sensations. Each depended on the
excitation of a specific receptor and remained independent of the others
until the process of perception was accomplished. This was a reminder
that in physics one could only speak of the vibration frequency of light,
whereas color is above all a psychological phenomenon. When a homog-
enous light like that of the sun fell on the retina, the perceived color was
approximately white because the three receptors were stimulated
together and equally intensely. But what happened when light of a single
frequency struck the retina? Young had already suggested that each of
the three receptors reacted to all wavelengths but with more or less
affinity. For example, red receptors reacted to all wavelengths but with
a maximum for the longest (i.e., the lowest frequencies), whereas blue
receptors preferred short wavelengths (i.e., the highest frequencies).
Green receptors preferred middle wavelengths and frequencies. So, for
a light that preferentially excited red receptors, green and blue were also

excited, but less so: moderately for green and weakly for blue. So this light was seen as red, but slightly unsaturated, because it contained a certain amount of white because of the simultaneous stimulation of other receptors.

As we often see in science, a discovery is rarely the work of a single person: An old problem has often undergone numerous pertinent attempts at an explanation, and frequently several researchers arrive at the same conclusion at the same time, sometimes without being aware of it. The story of color confirms this rule but also illustrates the preeminent role played by men such as Helmholtz, who combined technical competence with a rigorous theoretical approach that permitted an ambitious and durable synthesis, still highly respected today.

Contrast: Innate or Acquired?

We saw earlier the role of residual images and contrast in the genesis of Goethe's ideas, and it is therefore not surprising that Helmholtz devoted several pages to the subject in his *Physiological Optics*. His hypothesis about residual images depended on the notion of fatigue of retinal elements excited by light. In the story of the chamber maid in the scarlet bodice, Goethe was surprised after she left the room to see her silhouette against the white wall, but the bodice had turned green. For the physiologist, the explanation was simple: The retinal red receptors were fatigued and could not react for several minutes to any fresh light stimulus. When he looked at the white wall,the red receptors were therefore out of action, but those for green and blue had not been used so far and were thus active and, by subtraction (white minus scarlet), gave a subjective impression of the complementary color, sea green.

The problem of contrast was, however, much more difficult. The more his ideas were in conflict with the accepted doctrine of the time, and the more crucial a given point was for his theory of perception, the more pages he devoted to the problem in his treatise. First he warned his readers about the parasitic effect of uncontrolled eye movements that might give a false impression of simultaneous contrast. Indeed, if the eye fixated a red object and then a white background near the object, the retina was fatigued by the red, and it was not surprising that the wall appeared greenish. This was easily explicable by the mechanisms of residual images described previously and had nothing to do with simultaneous contrast, which was obtained with the eye immobilized and was not a purely sensory phenomenon as in the example just given. On the contrary, it was a process of perception, in which there was judgment and modification of appreciation without alteration of sensation.[35] Someone of average height appeared big beside someone who was smaller but looked small beside a taller person: For the same sensation of someone of moderate height, the perception was thus different depending on the size of his neighbor. For Helmholtz, this example illustrated two important rules of perception. First, one always tended to exaggerate perceived differences. Second, in the presence of a difference, whether of size or color, one unconsciously searched one's memory for a size or color as a reference to compare the two different elements.

It was the same for simultaneous contrast. If one juxtaposed two colors, one of which was highly unsaturated, one tended to accept the latter as "white" for reference purposes. For example, if one illuminated a sheet of white paper using two light sources—the white of moonlight and the orange light of a candle—one would not hesitate to see the paper as intensely white even if it contained a hint of orange that disturbed the equilibrium of the three fundamental colors of white. Then, if one placed an object between the candle and the paper, its shadow on the paper appeared bluish, although the paper on which the shadow was cast was as purely white as the incident moonlight.[36] This was explained by the fact that the shadow was a white that, unconsciously to the observer, contained too much orange, and only a blue color could reestablish its trichromatic balance.

For Helmholtz, simultaneous contrast was clearly psychological in nature because for identical retinal sensations, as determined objectively in terms of a physical light stimulus, the perception of colors could vary as a function of different unconscious criteria of appreciation. This thesis could doubtless be explained by Helmholtz's desire to introduce color perception into the wider context of his famous theory of perception but also by his hostility toward any suggestion that perceptive processes were innate rather than learned. It should be said that, at that time, there was no evidence for any "horizontal" connections in the retina, with functional implications for the retinal projection to the brain that might have played a role in the genesis of contrast.[37] But that is another story.

Helmholtz and the Painters

The problem of simultaneous contrast was of great interest to the painters of the nineteenth century, for whom Chevreul's book[38] was almost bedside reading. However, Helmholtz's research and that of Maxwell were only revealed to them rather late and indirectly, thanks to the writings of an American physicist and amateur painter, Ogden Rood.[39] In particular, the neo-impressionists Georges Seurat[40] and Paul Signac,[41] famous artists in the pointillist style, derived much inspiration from Helmholtz's results and the advice of Rood. They decided to avoid as much as possible mixing colors on their palette, using instead only spots of pure, luminous color placed side by side on their canvas. Due to the contrasts between these spots, the eye of the observer was presented with a complex play of complementary colors. Their fundamental colors were those of Helmholtz—red, green, and blue—and not those of Chevreul, which were also those of Goethe—red, yellow, and blue. Seurat rejected the traditional series of complementary pairs red/green, yellow/violet, and blue/orange and adopted Helmholtz's red/blue-green, purple/green, yellow/ultramarine, yellow-green/violet, and orange/cyan.[42] We must remember that Helmholtz, unlike Newton, distinguished between colored light that allowed "additive" mixtures and colored pigment that only allowed subtractive mixtures. Seurat believed that his spots of color had a luminosity close to that of natural light, and so their mixture was additive and the fundamental colors were those of Helmholtz.

In a lecture that he devoted to painting and in which he gave advice to artists, Helmholtz often cited Chevreul, whom he appreciated, but whose opinions he did not always share. To paint red flowers on a background of greenery, for example, assumed that one took into account the contrast created by the juxtaposition of the red and green fields. In fact, the eye saw the reds as redder and the greens as greener, especially at the borders between the two colored fields, which could be explained by the fact that the red field added its complementary color to the neighboring green field, and thus some extra green, and the red field became redder for the same reason. For Chevreul, the artist should not paint contrasts as he saw them. He must be content to reproduce colors with the intensity they had in the object being painted because it was the eye that was at the origin of the phenomenon. "To imitate faithfully the model," he said, "one must do differently from what one sees": If the painter knew how to create the necessary conditions, the effect leapt out spontaneously.[43] For Helmholtz, however, if the colors in a painting were as clear and brilliant as in real objects, contrasts would produce themselves. But precise measurements of luminosity made him state that, in most cases, the painter must reproduce contrasts as he saw them because one could not expect in a painting the reproduction of the living effects of contrast as one saw them in nature in brightly lit colored objects. One must reproduce objectively in the painting subjective visual phenomena because the scale of colors and luminosity in a painting could not reflect reality.[44]

9 The Founding Regard

I was told that I find what I am not looking for, while Helmholtz only finds what he is looking for.
—Claude Bernard[1]

Continuing his research in Königsberg, and now dean of the faculty, Helmholtz nevertheless made many journeys, notably to England, where he met Michael Faraday and Charles Wheatstone. He was dazzled by the extent of the cultural richness of London and the opulence of its museums, a sort of Babylon alongside which Berlin seemed to him no more than a village. He was in admiration before Westminster Abbey: "Therein professors of physics and chemistry lie next to kings."[2] These years could have been very happy for the couple, who then had two children, Katherina and Richard, if Olga's health had not in the meantime become more and more fragile.

In 1855, she could no longer tolerate the harsh climate of the Baltic, and her frequent lung problems prompted Helmholtz to apply for a chair in Bonn, where he was appointed in 1858 thanks to the support of his mentor Humboldt. He resumed his work on vision with renewed enthusiasm and even launched himself into new research on the physiological theory of music, which we discuss in a later chapter. But after three years on the banks of the Rhine, he moved with his family to Heidelberg, where the famous university promised him an institute of physiology that would live up to his needs. Heidelberg was, after Vienna and Prague, the oldest university of the Holy Roman Empire, and its cultural prestige was enormous. He had been tempted there by two friends, both influential professors in Heidelberg, Robert Wilhelm Bunsen in chemistry and Gustav Robert Kirchhoff in physics. They had discovered chemical spectral analysis, thus enabling them to better understand a problem that was fascinating in Helmholtz's eyes—that of the link between light and heat.

Scarcely was he settled in when he had to face the deaths of the two people who were closest to him, first his father in June 1859 and then his beloved wife Olga, a few days after Christmas, the same year. She was buried in the cemetery at Heidelberg, and her husband had engraved on the tomb: "Blessed be the rich seed of love that is sown here below."[3] The experience was painful: He remained in shock for several months and only recovered by maintaining an intense intellectual and musical activity. However, a year later, he met a young woman who had everything to please him, and especially to understand him and help him in his research career, Anna von Mohl. She was from a cultured and respectable Heidelberg family. Her father, a jurist in public law, was a member of the ephemeral Frankfurt parliament, and her uncle was professor of Persian at the Collège de France, which enabled her to live in Paris and perfect her French. They soon married and a new life began for Helmholtz, who found happiness again and devoted himself more than ever to research.

However, something had changed in him. Ten years later, his motivation had altered, and his taste for physics had gained the upper hand over physiology. He had gained considerable importance in academic and especially political circles. In 1870, he was appointed to the chair of physics of Magnus in Berlin. There he found his old friend du Bois-Reymond and devoted himself almost exclusively to the study of thermodynamics and electricity. He became rector of the university in 1877 and received a peerage a few years later. He ended his career by founding with his friend Werner von Siemens, and encouraged by the emperor, a physicotechnical institute devoted to fundamental and applied research. Siemens, a symbol of Prussian industrial might, was a family friend because his son, Arnold, had married Ellen von Helmholtz, Helmholtz's daughter by his second marriage. Siemens financed a major part of the new institute, a spectacular example of collaboration between private and public sectors. Helmholtz was its first director, which earned him the title of *Imperial Chancellor of Science*. He died in full activity in 1894 respected and honored by his peers.

The story is far from finished by this abrupt end. The biographical short-cut may be disconcerting for the reader, but it is indispensable for the description of Helmholtz's work in vision and audition, theoretically completed before his move to Berlin, but in fact under constant revision until the death of their author, so revealing a degree of intellectual evolution.

Helmholtz and Philosophy

The publication of his experimental results was always of prime importance to him and followed a well-established strategy. His priority was for his peers, in the shape of communications to learned societies or detailed papers in influential scientific journals. For nonspecialized scientists, physicians, ophthalmologists, and the aware public, he published various treatises that were often reedited and ensured his fame. Finally, he gave many lectures all over Europe, often accompanied with demonstrations designed for the general public. He attached much importance to these lectures and declared that he needed real artistic talent to give a lively but concrete presentation of scientific facts, without neglecting the more abstract considerations that helped understand them.[4] What was the place of philosophy in his scientific publications? In the nineteenth century, the rupture between science and philosophy that was already apparent in Newton and Galileo was consummated. Kant was perhaps one of the last philosophers to have accomplished some real scientific work when he published his research, based on Newtonian gravity, on the birth and evolution of the cosmos. It is therefore not surprising that in his scientific publications aimed at his peers, Helmholtz never resorted to philosophical arguments because he really did not need to do so.

Nevertheless, in the third edition of his *Physiological Optics*, he wrote an introduction that stood as a clear manifesto in favor of empiricism in science and in which he used philosophical concepts perhaps to better distance himself from them. Also, in his many lectures for the lay public, he adopted a much subtler stance, rejecting firmly anything that had an odor of metaphysics but coming to a compromise over "good philosophy." Among these "good" elements were the English empirical philosophers, with whom he felt close affinities. Equally, Kant and even Fichte, who had given some thought to the problems of perception and held views that were compatible with empirical science. Helmholtz continued what Kant had never had the intention to add to our wealth of knowledge by thought alone because he believed all understanding of reality was necessarily derived from experience, and the role of the philosopher must be restricted to looking for sources and the degree of justification of knowledge. However, he absolutely rejected philosophers after Kant's death, such as Schelling and Hegel, who, respectively, dominated the south and north of Germany because they believed that the acquisition of scientific knowledge was conceivable simply by their own thought, although it was in fact only possible through experience.[5] As to metaphysics, he was less conciliatory. In 1875, he said in a letter to his friend Fick[6] that, "To elaborate metaphysical hypotheses is like fencing in vain against a mirror," as we mentioned in chapter 5. At the end of his life, he also said that metaphysics played the same role in philosophy as astrology in astronomy. Many dilettante scientists were drawn to it hoping that it would help them acquire, quickly and easily, an understanding of the human mind and the future of the world.[7] For the anecdote, one might add that Helmholtz liked shows of conjuring and magic that he attended with his friends, deriving much pleasure from discovering the tricks used to deceive credulous spectators.

At the end of his *Physiological Optics*, he set precise limits for the natural philosopher who should not favor a metaphysical option to compensate for insufficient knowledge of psychic phenomena because he "must stick to the facts and try to find out their laws ... materialism, it should be remembered, is just as much a metaphysical speculation or hypothesis as idealism, and therefore it has no right to decide about matters of fact in natural philosophy except on a basis of facts."[8] Maybe while writing these lines the author regretted somewhat the audacious manifesto of 1847 (see chapter 5) in which with his young colleagues he extolled a radical materialist approach to scientific method. However, twenty years later, circumstances had obviously changed. Enveloped in a very strict political environment, conformist in its view of religion, Helmholtz certainly needed to project the right social image, but nothing suggests that he spoke against his own convictions in confronting materialism and spirituality because in the meanwhile he had discovered the importance of psychology and the impossibility of reducing it to physicochemical terms.

The *Handbook of Physiological Optics*

Helmholtz's monumental work of more than 1,000 pages is a systematic anthology of all knowledge about vision available at the time. Its genesis was due to a suggestion by one of his friends, Gustav Karsten,[9] who had begun an *Encyclopedia of Physics*, of which the ninth volume was to be a treatise on optics. The production of *Physiological Optics* was a long-term enterprise, even more so because its backbone was essentially the author's own experiments, and he was not always able to maintain the publication rhythm that the editor desired.

The first part was published in 1856, immediately after the author's arrival in Bonn. Dealing with the dioptrics of the eye, it was dominated by results obtained using the ophthalmoscope and especially the *ophthalmometer* (figure 9.1), a wonderful instrument invented by Helmholtz to make minute measurements of the eye. It enabled him to measure the curvature of the cornea and the variations of distance between it and the iris during accommodation. The second part was only published in 1860, in Heidelberg, and dealt with visual responses to light, simple and composite colors, intensity and duration of visual sensation, and residual images and contrast. The third part, in 1866, dealt essentially with perception of space.

The two first parts constituted an invaluable source of scientific fact derived from research by Helmholtz and his contemporary, or earlier, colleagues. The author clearly derived great satisfaction from using his vast knowledge of physics, geometry, and mathematics. A detailed study of these two parts is of essence highly technical and is outside the scope of the present book. However, the third part is more relevant because it related to Helmholtz's options in terms of scientific methodology and demonstrated the author's new major intellectual trend toward psychology that had become indispensable for his study of perception.

It began with a long introduction that the author rewrote several times throughout his lifetime, less to update it because of advances in science than to justify his theoretical opinions and refute the objections of his detractors. He developed a general theory of perception in which movement played an essential role. After several chapters devoted to, respectively, eye movements, perception of the direction of gaze and distance, and binocular vision, Helmholtz ended his work with a rather polemical text, in which he presented himself as a crusader for empiricism against innateness and settled some personal scores, notably with his colleague Karl Ewald Hering,[10] the successor of Jan Evangelista Purkinje at the University of Prague.[11]

Sensory Perception

Helmholtz launched the basis of his theory of perception even in his inaugural lecture in Königsberg.[12] He progressively developed it to become the essential of the introduction to his *Physiological Optics* and the subject of an important lecture in Berlin.[13] In the later versions of the introduction, his style became more and more difficult and even obscure.I It became more and more philosophical to ensure that his ideas

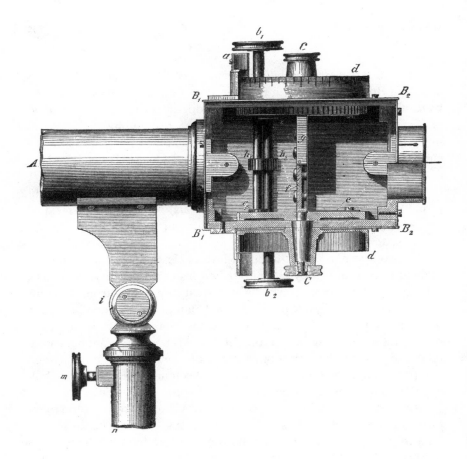

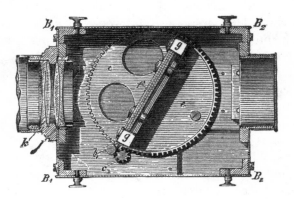

Figure 9.1
Helmholtz's ophthalmometer, which was used to measure many physical parameters of the eye, in particular the curvature of the cornea and the lens, was based on a micrometric system that allowed him to modify the angle of two glass plates through which he observed the surface of the eye obliquely. This caused a displacement of the observed image as a function of its position. In this way, it was possible to determine the shape of the surface. (Helmholtz 1856, I)

were coherent and lasting because he was aware of the changing mentality around him in a Germany where social unrest and mounting nationalism were distractions from his rational thoughts on perception and the unconscious.

Helmholtz was manifestly a remote disciple of the English empirical philosophers, notably John Locke, whom he cited several times. However, Kant was his principal guide in science because he had adopted Locke's ideas, as well as those of Hume, that were sometimes nihilist to impose a theoretical framework for experimental results that opened the way to modern rationality and scientific research. In developing his theory of perception, Helmholtz the empiricist respected, at least at first, Kant's doctrine according to which, to handle sensory information, the human mind disposed of a priori principles of space, time, and causality. Later he returned to these concepts and improved them by introducing movement into his physiology of perception, which was clearly no longer compatible with Kant's ideas. At a third stage, later in his life, he again became a keen partisan of Kant because it enabled him to give a noumenal unity—that is to say, an intelligible reality as opposed to a sensory reality—to psychology by paradoxically reintegrating in Kant's theory the space, time, and causality legitimized by physiology. Helmholtz had definitively enriched Kant's philosophy.

Specific Nervous Energy of Johannes Müller

The question of vision began in Helmholtz's inaugural lecture by a general study of the relationship between sensation and a detected object. The first principle was that what was sensed as light was not always light. The proof was that one experienced a brief light stimulus, like a flash, when one compressed the eyeball or passed an electrical current in it. The second principle was that part of the light spectrum remained invisible, notably the radiation that signaled heat and occurred beyond the visible red of the spectrum. Helmholtz accepted that light from an incandescent body or from the sun was made up of a large number of waves of similar physical nature but of different frequencies, all of which transmitted a greater or lesser amount of heat. For some of these frequencies, the light was visible to the eye and also warm to the skin, whereas others were invisible but nevertheless transmitted heat. Thus, said Helmholtz, one could conclude that there were sensations of light not produced by light, and there was light that did not evoke any sensation of light but rather of heat. The objective significance of visual sensation could not therefore be explained by an exact match between light and the sensation of light.

One was thus forced to accept Müller's conclusions, he continued, for whom the specificity of the sensation of light did not depend on any particular property of light but on specific activity in the optic nerve, stimulation of which could only give rise to a sensation of light. This rule also applied to other sense organs. For example, the skin felt perfectly well the warmth of a stove or the sun but was unable to sense light.

The theory of specific nervous energy[14] to which Helmholtz referred should be recalled here because it bore witness to Müller's difficulties in gaining acceptance for truly empirical physiology in early nineteenth-century Germany. The term "energy" intentionally did not signify anything in classic physics: It was no more than a principle that had never been refuted by repeated observation of the effects of visual stimulation, and that provided no information on the nature of retinal mechanisms. For Müller, one had to accept the axioms according to which the energies of light and dark and color were inherent not to external objects but to the very visual substance. This visual substance could not react if it was not in a state of responsiveness to the innate energy of light, dark, and color. For vision, light, dark, and color did not preexist like an external force that produced a sensation immediately. The visual substance was changed by all stimuli of whatever nature from a state of rest to one of excitation, and it made this excitation felt by its own action in terms of the energy of light, dark, and color. Müller's concept expressed in the form of an axiom, his conviction that the visual substance, which he did not want to define in anatomical, physiological, or chemical terms, was both a passive and active interface between an external stimulus, usually light but not necessarily so, and the sensation it generated thanks to innate energy of light, dark, and color. Visual sensation was thus identical to the specific energy of the visual sense.

The term "energy" should be understood in the context of the Aristotelian concept of ενεργεΙα made up of εν and εργονυ—"force inside"—and not in the modern thermodynamic sense. Müller's use of this term reminds us that between 1820 and 1830, Aristotle's doctrine still constituted an important arm for German physiologists who wanted to prioritize observation and experimentation and so opposed the power of natural philosophy. By neglecting to cite his sources, Müller was relying on general knowledge that all medical students of the time, who knew Kant and Aristotle almost by heart, should have had. As we discussed earlier, Helmholtz accepted the conclusions of his mentor, but not without reserve, because he was convinced that understanding sprang from experience, whereas for Müller specific energy was obviously innate. Müller's specific energy was nothing other than what Helmholtz called sensation. But how did he progress from sensation to perception?

Perception and Unconscious Inferences

Perceptions of external objects, he said in his *Physiological Optics*, were representations that were always the result of psychic activity. Therefore, the study of perception belonged to psychology insofar as it dealt with research into the nature and laws of the involvement of the mind in the production of perceptions. The results of psychic activity allowing one to recognize an object, whatever sense organ was involved, were analogous to the results of a judgment in which a cause was attributed to sensory stimulation. This judgment, which Helmholtz called induction or "sensory inference," was of necessity unconscious because it imposed itself irresistibly on one's consciousness without any room for free will. According to Helmholtz, the unconscious, already present in Leibniz's "weak perceptions,"[15] was really the anticipation of the future cognitive unconscious that obviously had little to do with William Carpenter's "unconscious cerebration"[16] or Freud's "subconscious," even if the latter was probably aware of it. Indeed, Freud worked in Vienna with the physiologist Brücke, student of Müller and friend of Helmholtz.

Leibniz stated, "Theophilus: We could perhaps add that animals have perception but not necessarily thought, that is to say reflection or what might constitute the object of reflection. We also have weak perceptions ourselves that we do not notice in our present state. It is true that we could very well notice them and reflect on them if we were not deterred by their multitude, which distracts us, or if they were not dispersed or obscured by stronger perceptions."[17]

Although unconscious, sensory inference obeyed the laws of logic, which Leibniz believed to be inherent in nature: a form of logic based on association of a given sensation with one's concept of a given object, which relied on previously acquired experience. Helmholtz said that reasoning happened by unconscious processes of association of ideas that resided in the unexplored parts of our mind, and the conclusions were like the results of inductive reasoning that took place without our knowledge. This was an empirical approach, in which conscious sensations were signs interpreted by our intelligence. Helmholtz thus opposed, not without creating a polemic, the nativist or inneist approach defended by several of his contemporaries, notable Hering, who supported a mechanical link using preexisting organic structures rather than a link through experience. Indeed Helmholtz, convinced by Kant's empiricism, could not tolerate the underlying concept in inneism of a preestablished harmony between nature and the human mind.

The logical consequence of Helmholtz's empirical approach, already developed in his *Habilitation* lecture,[18] was that the sensations of light and color were only symbols of reality, comparable to the symbols of human language that were in themselves purely arbitrary. At the end of his lecture, he stated lyrically that we should thank our senses, which miraculously gave us light and color as responses to particular vibrations and odor and taste from chemical stimuli. We should thank the symbols by which our senses informed us of the outside world for the spellbinding richness and the living freshness of the sensory world.

The Eye and Reason: The Birth of Gaze

It was one thing to recognize an object, but to locate it in space was quite another.[19] As we saw earlier, Kant seemed to have found a solution by postulating principles that the mind possessed a formal framework that allowed the construction of sensory patterns even before any perceptual experience. These are the famous transcendental or a priori principles of time and space, as well as the formal principle of causality resulting by deduction from the first two. Kant certainly did not believe that these principles could explain visual "localization" in space, which was psychological, but he was convinced that the acceptance of these principles justified experimental physiological or psychological research and opened up the field. However, two eminent scientists complicated this field, highly promising as it was, and so arose one of the most spectacular and abiding controversies of the nineteenth century.

First came Müller, in the wake of his theory on specific nervous energy, which he considered innate, such as the relationship of each point on the retina with external space. This space was projected like a two-dimensional image on the retinal surface, and every point in space was received by two identical or corresponding points on the two retinas. These corresponding points, he said, were always situated at the same distance from the retinal fixation point, close if the object was near the fixation point or farther away if the object was in the peripheral visual field. Further, an object in space must project to corresponding points if it was to be seen as a single element: if not, one would see double. Obviously Müller had no experimental proof for his theories, but his opinion was based on his anatomical study that demonstrated microscopically a partial crossing of nerve fibers in the optic chiasma. This suggested the

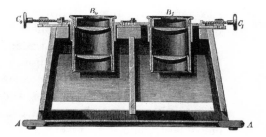

Figure 9.2
External aspect and section of a stereoscope constructed by Helmholtz and used in his experiments on three-dimensional vision.
(Helmholtz 1866, III)

possibility of topological organization, which spoke in favor of an innate character for visual localization in space. Unfortunately, Müller left many questions unanswered and did not explain how we see objects in space in three dimensions using binocular vision.

Later Charles Wheatstone relaunched the problem of binocular vision in depth with his discovery of the *stereoscope* in 1838 (figure 9.2). This instrument enabled him to see in three dimensions two flat drawings, or later two photographs, of the same scene, each presented simultaneously but separately to the two eyes. To obtain an impression of relief, the drawings had to be from slightly different angles of vision or the photographs taken with a camera having two objectives separated from each other by a distance equivalent to the distance between the two eyes. The two slightly different images were then fused when viewed binocularly, and the single scene was perceived in three dimensions. This discovery soon became the obsession of mundane salon society of the time, but we are more interested in its consequences. How was a third dimension created from two flat images? According to Wheatstone, depth vision was explained by the fact that the same point of the scene was projected by the double photographs on two nonconcordant points on the two retinae, and this disparity was at the origin of the depth effect and emergence of the third dimension. This new concept of depth, with the determinant role of binocular vision in producing it and the unexpected fact that stimulation of two nonconcordant retinal points could result in the perception of a single image, had not been foreseen by Müller, who, like many others, explained the perception of the third dimension by commonsense psychological factors, such as the masking of a distant object by a nearer object or the knowledge of the size of a familiar object allowing one to estimate its distance from the size of the perceived image. If one knew the size of an elephant, one would obviously see it as being far away if it appeared small in a landscape. In the years that followed, up to 1880, binocular vision enjoyed considerable scientific popularity, and most contemporary physiologists joined the often bedlam-like debate. Turner estimated that, between 1855 and 1859, some 29% of scientific publication concerning the eye and vision dealt with stereoscopy, binocularity, and eye movements, 70% of which came from German authors.[20]

Obviously Helmholtz also intervened in the debate. In particular, he undertook important research on the "horopter," the volume containing all points in space seen as single by binocular vision, and he concluded that his results cast doubt on his mentor Müller's theory of concordant points. But he went further, claiming that binocular fusion was an act of psychic inference, in no way innate, and unconscious judgment based on seeing a familiar object in binocular perspective. Thus, Helmholtz placed himself in the center of the controversy about depth perception. What was his intellectual approach in this fundamental debate between inneists and empiricists? Starting from an empirical point of view, he found a rather unexpected source of inspiration in the contemporary work of the metaphysician Lotze, who accepted the hypothesis that the mind gave rise to the perception of space by interpreting the sensations.[21] Lotze occupied a unique and original place in the academic world. Professor of philosophy at the Universities of Berlin and Göttingen, he was interested in psychology and physiology, with metaphysics as the starting point. It is not surprising that his position was uncomfortable, both in relation to his colleagues in his discipline, who saw in him a degree of antiphilosophy, and with natural scientists, who practiced experimentation rather than deductive reasoning. Nevertheless, his influence was important for the development of psychophysics, as Helmholtz rightly saw.

For Lotze, sensations from sense organs such as the eye differed from each other by a factor proportional to the frequency of movements of chemical particles depending on the stimulus. In practice, a sensation was defined exclusively by a frequency, different from one part of the retina to another. It would not have a physical dimension, a necessary condition for an immaterial mind, which obviously had no physical dimension. So the mind would have some general intuition about space that would make it not only capable of applying, but obliged to apply, this notion to the contents of sensations. It would do it by calculating the distance between light points in the visual field measured by the amplitude of the displacement of gaze needed to fixate them successively. What Lotze called "local sign" was the special quality that differentiated individual impressions of intensity, and that allowed one to situate their origin in space thanks to an always identical quantifiable eye movement, either real or even virtual if there was simply the intention to make the movement.

This theory of local signs was for Helmholtz a decisive step toward an empirical concept of spatial perception. He took it up and developed its principal ideas, avoiding certain contradictions and Lotze's often obscure phraseology. Spatial perception, he stated, was built up like perception in general from unconscious inferences, where one's judgment took into account both the sensory stimulus and its spatial localization derived from exploratory movements. Retinotopy, for example, was in no way innate, as Müller believed, but was the exclusive result of unconscious associations.

To avoid the impasse of Locke's and Hume's radical empiricism, Helmholtz had accepted as an indispensable presupposition to his research the Kantian principles of time, space, and causality. However, he imposed a physiological validity by demonstrating that these principles were verified and confirmed experimentally in the light of voluntary movement, which, he claimed, added a substantial richness to Kant's views (figure 9.3). Voluntary movement reinforced a subject's conviction obtained by unconscious inference because it allowed him to consciously confirm the cause of his sensations and so legitimize his perception. In fact, the eye was not the only organ to have this function. Other types of movement—of the body or the hand—collaborated in the exploratory function of the eyes, and this enabled the subject to have a representation of objects in space and record within his inner concept of time his memory of the progress of external events.

Two of Helmholtz's experiments were particularly revealing and deserve mention here on account of their simplicity and credibility. The first illustrated the importance of gaze in the everyday management of our sensory space. By common experience, if we changed our direction of gaze, it did not cause a change in our visual environment, which remained fixed in relation to our body. In contrast, a passive, rather than an active, movement of the eye, which was easily produced by closing one eye and pressing the other with a finger, caused a spectacular rotation of the visual field that appeared to slip sideways. Sensation from muscles could not be invoked here because in a patient with complete paralysis

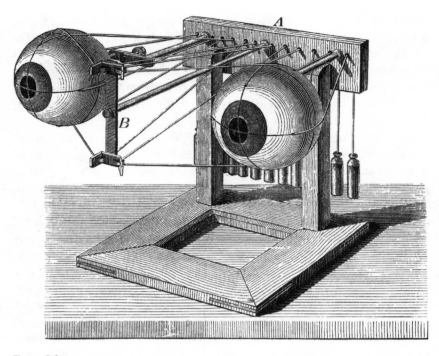

Figure 9.3
The "ophthalmotrope," a model used to study the basic mechanisms of eye movements: a
version built in 1857 by the German ophthalmologist Christian Theodor Ruete.
(Helmholtz 1866, III)

of the eye muscles caused by compression of the oculomotor nerves
inside the skull, the very fact of wanting to move the eye caused rotation
of the visual field in the same way, as if the desire to change the direction
of gaze unsuccessfully anticipated the impossible eye movement that
would have preserved the effect of a fixed spatial environment. Thus,
gaze always played an essential role in the correct perception of space.

The second experiment demonstrated that the delicate adjustment of
gaze in external space could be modified by experience. When a subject
looked at an object through prisms that deviated light rays by 16 to 18
degrees to the left, he saw the object more to the left than it really was.
If he was then asked to grasp the object with his eyes close, it was not
surprising that he aimed his hand too far to the left. Next, he was asked
to repeat the maneuver several times, but with his eyes open, until he
was able to perform the task correctly. After moving the object with the
subject still watching, the experiment of grasping for it with the eyes

closed was repeated, and then his hand reached the object without difficulty. It all seemed as if, by exploring with the hand, the subject had
learned where the object was in space and so was able to modify his
direction of gaze. So the eye was able to continuously "calibrate" its
relations with visual space by taking into account, at the level of perceptual processes, information about the spatial environment derived from
other sense organs. Helmholtz attached much importance to this sort of
experiment, which demonstrated the crucial role of continuous adjustment of gaze with experience and provided him with an important argument in support of his empirical view of physiology.

In the introduction to his *Physiological Optics*, Helmholtz emphasized
that it was in the nature of consciousness that ideas could not resemble
objects, that there was no preexisting harmony between the laws of
thought and those of nature, and that nature and the mind were not
identical. To defend such ideas was to align oneself with natural philosophy and reductionist materialism, two doctrines that were remote from
scientific rigor. Helmholtz had thus profoundly amended the ideas of his
youth and his philosophy of explaining biology exclusively on basic
physicochemical forces as he had defined them in the manifesto of 1847.
Pragmatically, he had discovered the importance of psychology not being
reducible to physiology, necessary to give some sense to the latter. For
its author, this was doubtless less a contradiction than a fruitful episte-
mological rupture that ensured physiological coherence for his work. He
said that to test and confirm the legitimacy of one's perceptions by a
voluntary movement was the product of free will derived from reflexive
consciousness and the ego. These voluntary movements obeyed different
laws from those that one observed in the outside world, where there only
existed obligatory effects in response to adequate causes, and they were
the only means for the ego to become aware of the chasm between itself
and nature.[22] Here Helmholtz was close to the philosopher Fichte, for
whom action was not a property of the ego but its substance.

Surprisingly, in his experiments, Helmholtz consistently ignored anatomical data on the nervous system and strangely never referred to the
brain. The reason for this omission was probably his mistrust of the
concept of an anatomophysiological correlation that might have suggested that anatomical and psychological processes were identical, which
was exactly the battle cry of the natural philosophers.

10 For or Against Pythagoras?

Beauty is a manifestation of secret natural laws, which otherwise would have been hidden from us forever.
—Johann Wolfgang Goethe[1]

Hearing Research: Why and How?

Helmholtz's research on hearing was audacious and of great scientific significance. It still evokes astonishment and respect today because it represents a rare example in the history of science of such ambitious empirical research carried out by a single man in so short a time. In fact, the whole process was conducted rather haughtily. In a letter to his friend du Bois-Reymond in October 1855, when he had left Königsberg for Bonn, where he had been appointed Professor of Anatomy and Physiology, Helmholtz announced for apparently the first time his intention to begin research on audition. He asked his friend to have made for him a polyphonic siren that he needed for his experiments.[2] Only months later, in May 1856, he wrote again to du Bois-Reymond to report the first results of his research, which he communicated immediately to the Berlin Academy and published them without delay. In another letter in May 1856, but this time to his colleague Wilhelm Heinrich Wittich, a microscopist at Königsberg, he expressed his firm desire to reduce the whole of the laws of musical harmony to the simplest possible basic elements and immediately proposed a scientific explanation of tonic consonance and dissonance.[3]

In a lengthy lecture in Bonn, "the home town of Beethoven, the greatest of the heroes of music,"[4] the following year, he gave an almost complete outline of his results, which much impressed his audience. Finally, in 1863, he published a weighty volume aimed at "readers from very varied educational backgrounds, pursuing very distinct interests," in

which he presented his experimental results in a coordinated, pedagogical fashion, placing them in their historic and artistic contexts.[5] Thus,
despite other academic activities, notably his continued studies on vision,
just a few years were enough for him to complete an ambitious research
project, undoubtedly one of the finest feathers in the cap of neurological
science in the nineteenth century.

There is perhaps no better testimony to the importance of this work
than the words of Georg von Békésy, the renowned master of auditory
physiology in his Nobel address of December 11, 1961: "For me, the most
stimulating book on hearing was Helmholtz's *Die Lehre von den Tonempfindungen*. ... Helmholtz's method of viewing physiology and psychology
in physical terms is today just as fresh as the day it was written. Helmholtz's magnificent start, however, was followed by stagnation in auditory research, and for almost 100 years the universities taught about the
same thing."[6]

Helmholtz's determination to tackle head on and so expeditiously
such a vast problem as the physiology of audition, from the physical
nature of the sound stimulus to musical esthetics, inevitably intrigues us.
What then was the motivation that inspired him so forcibly? It is quite
understandable that after studying the eye and vision, he would have
been attracted to turn his attention to the ear and audition even more
so because he was obviously aware of recent progress in physics and
probably knew about the controversy between the acoustic specialists
Georg Simon Ohm and August Seebeck concerning the pulsatile or wave
nature of sound. But there was probably more than these legitimate
motives to confirm for the ear his results for the eye or to take a stand
in the arguments about the nature of sound, in what was a rather oppor-
tune or indeed opportunistic manner. According to his introduction to
his *Sensations of Tone as a Physiological Basis for the Theory of Music*,
he endeavored to bring together across their common borders aspects
of science that had until then remained too isolated from each other—
namely, physical and physiological acoustics, music, and aesthetics. He
added that science, philosophy, and art were unreasonably far apart, and
this led to real difficulty in understanding the language, method, and
object of each of these approaches. So his project was clearly to unite
within a single field science and aesthetics. In his introduction, he spelled
out his thoughts in a concrete fashion.

It may be useful here to mention that the intervals between two
musical notes are named as follows: C-D second, C-E third, C-F fourth,
C-G fifth, C-A sixth, C-B seventh, and C-C octave. Helmholtz recalled

that Pythagoras already knew that strings of similar nature, under the same tension but of different lengths, gave perfect consonance, octaves, fifths, and fourths if their lengths were in the ratio of 1 to 2, 2 to 3, and 3 to 4. But he asked, "What have consonant musical chords to do with the relationships of the first six whole numbers?"[7] To this question, many musicians, philosophers, and physicists had replied that the human mind had the faculty of appreciating the numerical relationships of sound vibrations and that it experienced a particular pleasure to discover simple, easily perceptible relationships. So did the consonant or discordant character of musical chords have something to do with a sort of natural order? But was it not better to look for an explanation of this mysterious harmony between whole numbers and musical consonance in the world of empirical physics?

We see later how, thanks to some impressive laboratory experiments, Helmholtz untied this Gordian knot in favor of the empirical hypothesis, although he demonstrated a certain ambiguity in the relationships that he saw between a global view of music and nature. On the one hand, he did not deny the power of a musician to create within an aesthetic order and admitted that "esthetic analysis of the greatest musical works, to understand the reasons behind their beauty, still almost always encounters apparently insurmountable obstacles." On the other hand, "beauty is subject to laws and rules related to the nature of human reason ... and we learn ... to recognize and admire in a work of art the image of a particular world order ruled throughout by law and reason."[8] As one can see, Helmholtz's intentions once again formed part of his comprehensive unifying project necessitating knowledge of physics, physiology, history, musicology, and even philosophy, not to mention prodigious skill in the use of laboratory apparatus. But, in fact, was this ambitious encyclopedist really a musician?

Helmholtz the Musician

Of course he was a musician! Were not music and philosophy the two mistresses of any self-respecting German intellectual? Helmholtz was no exception to this tradition. Indeed, his father had taken care to encourage in his son from a very early age a taste for practicing music within the family. The following examples clearly demonstrate his devotion to music throughout his life. At age seventeen, he left the family home to go to Berlin to study medicine. His piano went with him. He shared his room with another student, also a pianist, who, he told his parents, "plays the piano with frenetic skill, but only enjoys highly colorful pieces and the new Italian music." Ferdinand Helmholtz was rather worried and wrote to his son Hermann to encourage him to practice his music courageously and not let his friend play all the time in his place, simply for convenience, because he himself had lost the little that he knew in just that way, "but especially you must not allow your taste for profound, spiritual German or ancient music to be corrupted by glittering new Italian extravagance which flatters the ear: the second is seductive, but the former is edifying." Hermann replied to his father by return that he need not fear, that he did not like the modern music played by his companion, and that it incited him to play himself. In any case, he was rarely satisfied by the expression and style of another pianist, and it was by playing himself that he experienced the greatest pleasure! Indeed he played an hour a day, and more on Friday, Saturday, and Sunday. He was studying the sonatas of Mozart and Beethoven and working on the score of Glück's *Armide*.[9]

At an invitation by some friends of his father during his medical studies, he made the acquaintance of the composer August Schaffer, who played the overture of an opera that he had just completed, *Emma von Falkenstein*, which Hermann found quite good.[10] Later, on holiday at his uncle's in Königsberg, he had to be content with the musical scores that he found there: a single Mozart concerto, but also some excellent pieces by Johann Strauss the elder,[11] Joseph Lanner, Carl Czerny, Daniel Auber, and Vincenzo Bellini. He played them all one after the other but felt so disgusted at the end that he sought refuge in the Mozart concerto and the studies of Johann Baptist Cramer to feed his spiritual devotion.[12] From then on, at the age of seventeen, young Helmholtz's tastes seemed very traditional.

The years went by, and in 1847, and we have seen earlier, Hermann became the very serious Dr Helmholtz, an army physician in Potsdam, where he met a young girl, Olga von Velten, who sang quite charmingly. They often made music together, and he soon fell madly in love with her. Shortly after their engagement, he arranged with Olga to attend a symphony concert in Berlin. Unfortunately, she did not go, and he sent her a tearful message: "You did not come and my ability to listen was distracted. It was as if your soul, able to interiorize music so deeply, had always opened my spirit to the harmony of sound. My ears heard the succession of musical notes, but my mind remained desperately deaf. Of course they played a Mozart symphony, one of his most beautiful works, capable above all others of plunging me into ecstasy. But I was there quite alone, abandoned by the beautiful half of my soul, and it would have been the same if I had been listening to scales on the piano. It was not until the overture of *Coriolanus* that I regained my spirits: it is a pure jewel, short and concise, so proud and resolute, oscillating between nervous anxiety and the confusion of battle before finally dissolving into a few melancholic notes: a masterpiece of unequalled grandeur."[13]

Did this letter reveal a warm heart or rather a heart of steel, tinged with conventional sentiments of love, but needing music to reach true heights and overcome the hardships of life?

Several years later, when traveling back from Geneva to Heidelberg in response to an urgent call from his wife that their son Robert was seriously ill, he nevertheless stopped for a few hours in Freiburg-in-Breisgau to see again the organ that he had heard some years previously and that he much admired. "The organ is really remarkable," he wrote, "perhaps even more for an expert in acoustics than a musician. I must say that I had until then no idea of the effects of such an instrument, both in terms of its mass and power, as well as the multitude of its acoustic colors." In Paris in 1866, French colleagues from the Ecole Normale took him to visit the workshop of the great organ maker Aristide Cavaillé-Coll, as well as the famous organ, the largest in Europe, that he had conceived for the church of Saint Sulpice. He was full of admiration for the intelligence and creativity of the master.[14]

Soon after, at the famous concerts at the Conservatory, he heard a Haydn symphony, an extract from Beethoven's ballet *Prometheus*, *A Midsummer Night's Dream* by his beloved Mendelssohn, and choral music by Bach and Handel. "The choirs are better heard in Germany," he wrote to his wife, "but the perfection of the orchestra is unique in its field. The oboes chattered in the Haydn symphony like the gentle breath of a Zephyr; the chords were so right, especially the first high chords in Mendelssohn's overture, that even the first very high chords, with which it also finishes and that usually sound so wrong, were as pure as gold. The *Prometheus* was the most wonderful harmonic sound, with the horns in the foreground. This concert, after the *Venus de Milo*, was my second taste of perfect beauty."[15]

Several years later, in 1875, his father-in-law, to whom he was very close, died. With his children he took refuge by listening to soothing music. One evening, after playing Beethoven's quartet, opus 130, with his great friend Joseph Joachim as first violin, he wrote some movingly profound reflections: "Beethoven's opus 130, monstrously grandiose and serious, but deeply sad, has only today become totally transparent for me. Every bar of the adagio was played perfectly; it is like a tearful dream of lost ideals, and perhaps the archetype of Tristan dying for love, the impalpable wave of an infinite melody."[16]

A year later, husband and wife attended the formal opening of the theater and festival of Bayreuth—the place, as Lavignac said, where "you go as you please, on foot, on horseback, by coach, by bicycle, by train, but where the true pilgrim should go on his knees."[17] As was the case for the whole of the musical world, said Helmholtz's biographer Koenigsberger,[18] they were captivated with enthusiasm for the novelty and the grandiose nature of the works of Richard Wagner. At the twilight of his life, he returned to Bayreuth with his wife in order to enjoy once again "the incomparably beautiful *Mastersingers*." After his death, Cosima Wagner wrote to Anna von Mohl, Helmholtz's second wife, "With sadness I realize what a friend and protector I have lost. He who in spite of his important position extended compassion to me at a time that showed me no sympathy, and in a world that has remained without sympathy."[19] One might imagine that Cosima was alluding to the difficulties that Wagner's music had endured before being finally recognized, but that was not the case. The correspondence between Cosima Wagner and Anna von Mohl[20] informs us that Cosima's gratitude concerned the courage of the Helmholtz family in supporting and welcoming Wagner and Cosima on several occasions while she was still the wife of the conductor Hans von Bülow[21] and when her long relationship with the composer of *Lohengrin* was causing a great scandal.

Helmholtz was a musician, not only passively but actively. He was an enlightened amateur pianist, no more nor less perhaps than many German intellectuals of his time, but his technical knowledge of the functioning of musical instruments was considerable. Music was extremely important to him, and the prior examples show to what extent it sometimes helped him transcend problems of life and death. His taste was eclectic and refined. He esteemed, and perhaps liked, Wagner, but one suspects that his true taste was rather for much more traditionally oriented music. He was not indifferent to the emotional impact of music, and we have seen several examples of that, but he needed to channel his emotion and enclose it within a defined framework. It is doubtful that he preferred Wagner to Mozart or Beethoven, but it must have been difficult for such an honored and official scientist as he to openly admit it in imperial Prussia.

Musical Consonance: Pythagoras or Euler?

Musical instruments existed well before Pythagoras. He, however, made a brilliant discovery: the relationship between consonance and whole numbers. He used the *monocord*, an instrument with a single string stretched over a sound box. A bridge placed on the string divided it into two segments, each producing a sound of which the pitch depended on the length of that segment. What Pythagoras noticed was that for the two segments to produce sounds at consonant intervals—that were pleasing to the ear—it was necessary for the respective lengths of the two segments to be in the ratio of two small whole numbers. If he placed the bridge so as to leave on one side one third of the string, with two thirds on the other side, the two segments were in the ratio 2:1 and the sounds produced were an octave apart, the longer segment sounding the lower note. If two fifths of the string were on one side of the bridge, with three fifths on the other side, the ratio was 3:2 and the sounds formed a fifth. However, at that time, people were far from knowing that sounds were produced by vibrations that could be measured. Only later did research by Galileo, Newton, Leonhard Euler (1729), and Daniel Bernouilli (1771) on the movements of strings demonstrate that the relationship of their length was inversely related to the frequency of the vibrations forming the sounds. Just as for string length, the ratio of vibration frequency to achieve consonant intervals was expressed by small whole numbers.

Why did consonant intervals occur when the ratio of vibration frequency was expressed as small whole numbers? As Helmholtz pointed out, Pythagoreans relegated this mystery to the level of speculation about the harmony of the spheres, whereas Euler, a Swiss mathematician in the eighteenth century, judged that the human mind found particular satisfaction in simple relations. It was, however, difficult to imagine how the mind counted sound vibrations and derived pleasure from them when they were related by ratios of small whole numbers. The speculations of Pythagoras and Euler were therefore both rejected, and Helmholtz devoted considerable effort to solving this problem.

The Natural Scale

Before going further, we should try to follow Helmholtz in his journey to the heart of his experiments on sound. What was known in this field when he began his research?

Acoustics in the eighteenth century after Euler and Bernouilli was virtually moribund, and knowledge about the vibrations of strings was more useful for the development of theories about differential equations than for acoustics as such.[22] However, one should not neglect the excellent research of Ernst Chladni[23] on the patterns formed in relief in a fine layer of sand on a vibrating membrane that led, in 1802, to the first real monograph on experimental acoustics. It was only in 1825, after a quarter of a century of lack of interest, that the important work on waves by the brothers Ernst and Wilhelm Weber appeared, which introduced Germany to the mathematical works of French scientists such as Siméon Denis Poisson, Pierre Simon de Laplace, and Augustin Louis Cauchy. This research took a new turn after 1830 when the construction of sirens became just as important as the tuning fork for the study of sound. Charles Caignard de Latour (1827), Félix Savart (1830), and Friedrich Opelt (1834) produced several. Heinrich Wilhelm Dove (1851) and Helmholtz made numerous improvements to them.

When the latter began his work on acoustics in 1856, it had been at a standstill since 1843. In that year, a long and sometimes ferocious controversy between two eminent acousticians, Ohm and Seebeck, came to an end without a solution. At the origin of this controversy was the opinion of Seebeck that sound was not a wave but a series of pulses of arbitrary nature, whereas for Ohm sound consisted of sinusoidal waves. Later Seebeck admitted that sound could be a wave, but the controversy changed to the causes of the "timbre" of sound and the influence of its form on this.

In his book, Helmholtz noted first that vibrations produced by an acoustic source such as a musical instrument were transmitted to our ear through the air around us, such that the eardrum was subjected to alternating increases and decreases of pressure. This propagation from the source to the ear was by waves that progressed by a reciprocating forward and backward movement of air molecules. A point to recall is that the vibrations of a sound wave can be considered longitudinal, in the same direction as that of its propagation, whereas vibrations of light waves are perpendicular, like waves in the sea. The *intensity* of the sound was proportional to the *amplitude* of the vibration of the sound source and the *pitch* of the sound proportional to the *frequency* of vibration. As to the timbre of the musical sound, its causes were so controversial in Helmholtz's day that, before defining them, he was obliged to reconsider the basic problem. To do that, he needed to master quantitatively the techniques of sound production (figure 10.1).

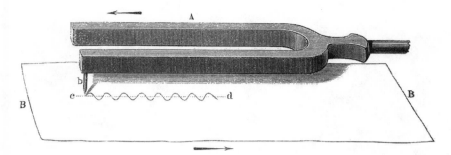

Figure 10.1
Drawing of a simple graphic instrument used by Helmholtz in his studies of acoustics to record on paper the vibrations of a tuning fork. The trace reproduced the form and frequency of the vibrations due to the relative movement of the paper beneath the tuning fork.
(Helmholtz 1863)

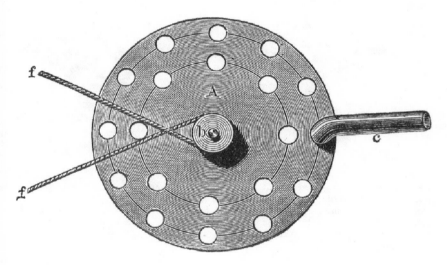

Figure 10.2
Disk with two concentric series of holes from one of the sirens used by Helmholtz. When the compressed air tube c was opposite the inner circle of holes, the frequency of the fundamental tone of the siren was reduced by a factor corresponding to the ratio of the number of holes (12/8, that is 3/2, or a fifth in musical terminology) independent of the speed of rotation.
(Helmholtz 1863)

The newly discovered siren was important as a sound generator in the hands of Seebeck and Ohm. Helmholtz had some built before moving from Königsberg to Bonn, and he made numerous improvements in them (figures 10.2 and 10.3). The principle of the siren is easy

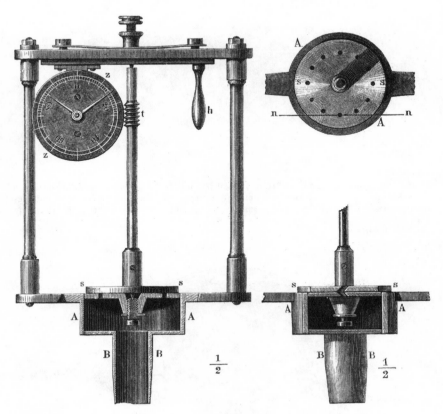

Figure 10.3
Cagniard de La Tour's siren modified by Helmholtz. The apparatus is seen from the side
and partially sectioned in the left and bottom right drawings. Air entered the chamber A-A
from the tube B-B and passed through the disk S-S (seen from above at top right). In this
case, the disk has twelve holes. The speed of rotation of the disk could be measured by
the counter visible in the left drawing.
(Helmholtz 1863)

to understand. The first part of Seebeck's siren was a disk turning around
its axis at a variable and measurable speed. The disk was pierced with
equally spaced holes in two concentric circles, twelve holes in an outer
and six in an inner. A fixed tube brought air under pressure and was
placed so that its orifice coincided exactly with a hole in one of the circles.
It is easy to see that if one turned the disk rapidly, the compressed air
escaped each time the orifice of the tube was opposite one of the holes,
but it could not escape when the orifice was opposite the surface of the
disk between two holes. So one obtained a succession of air pulses at a
frequency equal to the number of revolutions per second multiplied by

the number of holes through which compressed air passed. For example, if the disk turned 200 times per second and one was using the concentric circle with 12 holes, the frequency of the sound was 2,400 cycles per second.

It is easy to see the usefulness of this type of siren in which one could readily modify the speed of rotation of the disk, as well as the shape and number of holes. Taking the example of a disk with six and twelve holes described earlier, if one sent compressed air successively through the circle with six holes and then the one with twelve, if the disk speed was constant, the two sounds obtained were exactly an octave apart. If the disk speed was increased, each sound was higher in pitch, but the interval between them was unchanged at an octave. We can deduce, like Helmholtz, that a note an octave above another vibrates at exactly twice the frequency. If the disk had a series of twelve holes and another of eight, the sounds produced for a constant speed had an interval between them not of an octave but a fifth, and Helmholtz concluded that two sounds formed a fifth when the higher one vibrated three times while the lower vibrated twice.

Dove's polyphonic siren had four series of holes: eight, ten, twelve, and sixteen. The series of sixteen gave a sound an octave above that of eight and the fourth of that of twelve. The series of twelve gave the fifth of that of eight and the major third of that of ten. These results were known before Helmholtz, but he verified them meticulously. He succeeded in producing a regular and constant rotation in a siren by means of a small electromagnetic apparatus, in which the electrical current was interrupted by the centrifugal force of a weight as soon as the speed of rotation reached a certain limit. So the speed of the machine increased up to this limit but could never exceed it. A small turbine was adapted to the disk to force air through the holes (figure 10.4). Thanks to this system, the siren gave sounds of an extraordinarily constant pitch, almost as good as the best organ pipes.[24]

Helmholtz was able to establish for consonant intervals expressed in terms of pulses per second the same relationships, but of course inversed, as those discovered by Pythagoras in terms of string length, that is to say 2:1 for the octave, 3:2 for the fifth, 4:3 for the fourth, and 5:4 for the major third. There was nothing surprising in this because the vibration frequency was inversely proportional to the length of a string. By using sirens with different numbers of holes, the ratios 9:8 for the second interval, 5:3 for the sixth, and 15:8 for the seventh were established giving the following ratios for all the notes of the scale:

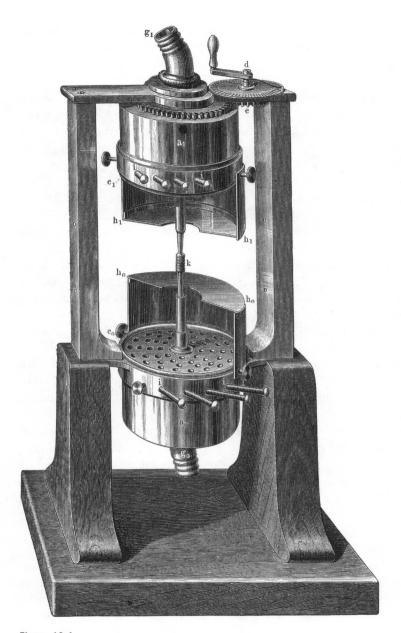

Figure 10.4
Double siren used by Helmholtz for the study of beats and for the production and analysis of resultant or combination sounds. The apparatus was based on the combination in a single instrument of two Dove polyphonic sirens.
(Helmholtz 1863)

C	D	E	F	G	A	B	C
1	9:8	5:4	4:3	3:2	5:3	15:8	2

This gave, for example, for a two-foot organ pipe the following scale expressed as frequency of vibration:

C	D	E	F	G	A	B	C
264	297	330	352	396	440	495	528
(1	9/8	5/4	4/3	3/2	5/3	15/8	2)

This constituted a natural scale, and we see later that it was slightly different from the scale that Pythagoras established, but especially from the so-called equal tempered scale used since the eighteenth century and best known from the famous *Well Tempered Clavier* by Johann Sebastian Bach. Helmholtz always maintained that this scale sounded false to the ear.

Harmonics and Musical Timbre

As Helmholtz reminded us, the shape of the wave carrying vibrations from the sound source to the ear played no role in the loudness of the sound, as long as the vibrations were fast and periodic. It could be sinusoidal, pulsatile, or even consist of successions of slowly rising phases followed by sharply falling phases, such as in the case of the bow of a violin being slowly lifted and then suddenly falling again on the string, like a hammer driven by a water wheel. In contrast, the timbre of a sound was dependent on the waveform. Although this factor was known to be important at the time, no one knew the mechanisms, and here Helmholtz played a defining role. He began with an old obser-vation little known to nonmusicians: When one listened with great care to the continuous sound of a violin or any other musical instrument, one could hear, in addition to the *fundamental* tone of which the pitch was determined by its frequency, a series of higher pitches that were called harmonics. Whatever the instrument, the harmonic series was the same:

First harmonic: the octave (C1) above the fundamental (C0), and thus twice its frequency

Second harmonic: the fifth (G1) of this octave, and thus three times the frequency of the fundamental (C0)

Third harmonic: the second octave above (C2), and thus four times the frequency of the fundamental (C0)

Fourth harmonic: the major third (E2) of this octave, and thus five times the frequency of the fundamental (C0)

Fifth harmonic: the fifth of this octave (G2), and thus six times the frequency of the fundamental (C0)

Then, with decreasing intensities, the harmonics at frequencies 7, 8, 9, 10 times, and so on, more than the fundamental.

One could conclude that a musical sound was produced by the superimposition of several elementary or *partial* tones, of which the lowest was the fundamental and the other, higher, notes the harmonics. The frequency of the harmonics was always a multiple of that of the fundamental, which recalled the relationships in terms of simple whole numbers already described in the experiments with sirens. For example, the fifth G-C (G1-C1) had a ratio of 3:2, and the major third E-C (E2-C2) had a ratio of 5:4. Having said that, if we play a note on a piano or listen to a sustained note on a clarinet, it is difficult for the untrained ear to distinguish clearly the harmonics of the fundamental. Even Helmholtz stated that he had needed much practice and effort before succeeding himself. What was more, to convince an unskilled observer it was not enough to ask him to describe what he heard, it was necessary to have objective arguments to undertake a quantitative study of harmonics, a necessary condition to separate Seebeck and Ohm.

After several attempts, he perfected an instrument, called a *resonator* (figure 10.5), to allow him to isolate a specific frequency from a mass of tones. It consisted of a hollow sphere or tube with two openings, one with straight edges and the other in the shape of a cone that could be placed in the ear. He had several resonators made, each with a specific resonant frequency, and this enabled him to hear, and to let others hear, the otherwise almost inaudible harmonics of a given fundamental tone. To hear clearly the first and second harmonics (C1 and G1) of the fundamental C0, for example, it sufficed to play this note and listen to it through resonators of which the frequency corresponded, respectively, to C1 and G1. Using several such resonators, Helmholtz conducted a systematic study of musical instruments and demonstrated that the characteristic timbre of a given instrument differed from others exclusively by the respective intensities of the different harmonics of the fundamental note. Helmholtz suspected therefore that the ear recognized the timbre of a sound by perceiving one by one the intensities of the harmonics superimposed

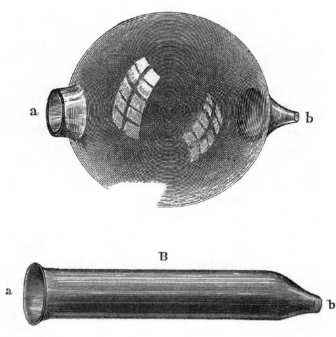

Figure 10.5
Two resonators used by Helmholtz to amplify the sensation produced by tones of specific frequencies. By applying the sound to be analyzed to the aperture a and the ear to b one heard distinctly the tone corresponding to the natural frequency of the particular resonator.
(Helmholtz 1863)

on the fundamental note. But before resolving this problem, he needed to return to that of the relationship between the shape of a sound wave and the timbre it produced, something that had remained obscure.

Fourier's Triumph

One must admit, as did Helmholtz in fact, that Ohm was the first to suggest that the ear was capable of analyzing sound. Indeed, he claimed that the airborne vibration of molecules according to the laws of the pendulum or the tuning fork could alone explain the sensation by the ear of a simple sound, and that a complex sound, with its accompaniment of harmonics, could be broken down into a summation of simple frequencies. We can obviously produce an enormous variety of wave forms if we compose a complex sound artificially by the superimposition of regular waves of different frequency. This is where the great French mathematician Joseph Fourier[25] intervened. He was well known to Ohm and Helmholtz, and he formulated the law that, "Any given regular periodic waveform can be composed from the sum of simple waves of which the frequencies are one, two, three, four etc times greater than the given frequency." Helmholtz formulated Fourier's law as applied to acoustics in these terms: "All vibratory movement of the air in the external ear corresponding to a musical sound, can ... be considered as the sum of a number of vibrations corresponding to partial tones of the sound."[26]

This strict analytical approach to complex sound, together with an objective approach to scarcely audible harmonics above the fundamental, thanks to resonators, sounded the death knell for Seebeck's ideas. Indeed, for him part of the intensity of the harmonics of a complex sound dissolved into that of the fundamental tone that it reinforced, and it was the little that remained which permitted the perception of harmonics. But without the help of resonators, it was impossible for Seebeck to hear the harmonics that Ohm's law required, and he wrongly assumed that their intensity reinforced that of the fundamental tone.

The way now seemed clear for Helmholtz to develop a theory of audition and perception of musical timbre. However, an essential question needed to be resolved: Did the ear only perceive the timbre of a complex sound by analyzing the different partial tones, thanks to a Fourier analysis, or did other characteristics of the complex sound also play a role? One might think of the more or less large phase difference between waves corresponding to the partial tones that might modify the timbre, defined earlier as the sum of the respective vibrations of the fundamental tone and its harmonics. This question was all the more important because to imagine a theory of audition in which the perception of musical timbre would require not only the detection of partial tones, but also their phase relations, would be too complex a process. Here one can admire once again the great scientist for providing an elegant solution to this problem, derived as much from his technical inventiveness as to his gifts as a theoretician and mathematician. His remarkable mathematical models were summarized in the annexes to his work and so are rarely noticed by his readers.

He constructed an absolutely original instrument that was able to generate continuous pure tones of sufficient intensity to be heard by several people at a time (figure 10.6). The generator consisted of a tuning fork that could only emit a weak tone of brief duration when struck once, but in this case the arms of the fork were vibrated by an electromagnet that successively attracted them when the current passed through the

Figure 10.6
Pure tone generator used by Helmholtz in his studies of musical acoustics. The instrument was based on the application to a tuning fork of an oscillating magnetic field obtained through the action of an alternating current passing through the coil of an electromagnet. Thus, the tuning fork oscillated continuously instead of gradually stopping, as it would do naturally.
(Helmholtz 1863)

magnet's coil and released them when the current was interrupted. Helmholtz's achievement was to have perfected an apparatus that enabled the tuning fork to vibrate continuously and automatically at its specific frequency. He achieved this by regularly making and breaking the electrical circuit by means of the fork's movements. The sound was pure because the frequency of vibration of the tuning fork was locked to its own inherent resonant frequency, and it was continuous as long as the operator left the electricity connected. The basic sound was, however, not very loud. Hence, to correct this, he installed alongside the fork a resonator of the same frequency. By using several generators of this type, Helmholtz was able to compose complex sounds with fundamentals and harmonics, and he was able to vary their timbre by changing the number of harmonics as his theory predicted. So he could reproduce artificially a number of typical timbres, which caused quite a stir at the time: "vowels of the human voice ... certain registers of the organ ... the nasal sound of the clarinet ... and the soft tones of the horn." However, the sound was far from perfect because it often lacked the characteristics of the real sound, such as "the sharp whistling sound of the current of air as it breaks against the lips of the organ pipe" and "the sharp upper harmonics" of reed instruments.[27]

These generators also enabled him, as was his initial objective, to modify the phase relationship between the sound that entered the resonator and that which came out simply by narrowing the orifice of its exit and thus reducing the resonance. For example, a reduction of the intensity of resonance of 10% was enough to put the sound coming from the tuning fork and that coming from the resonator out of phase by 35 degrees. In this way, Helmholtz was able to demonstrate that the musical timbre of a complex sound was in no way modified when one or more harmonics were out of phase. The conclusion was obvious: The theoretical physiologist expected the ear to be capable of analyzing a complex sound into fundamental and harmonics without taking into account phase relationships among frequencies, which were always in fact a multiple of the fundamental.

Helmholtz's Ear

The physiology of the ear at that time was of necessity only theoretical and based on contemporary anatomical knowledge. There was no hint of the spectacular technical progress that experimental neurophysiology has made today, through electron microscopy, microsurgery, and the study of single neurons with microelectrodes. Helmholtz's intuition was all the more remarkable for his implacable logical reasoning, the elements of which had been carefully and systematically put in place beforehand through calculation and experimentation. Helmholtz's studies on acoustics, of which we have just examined the essentials, clearly showed that the perception of the timbre of a complex sound relied of necessity on its reduction into its different parts, fundamental and harmonics. He asked us to suppose that we lift the dampers of a piano by using the right pedal and that we let a given note resonate at the soundboard. We would immediately hear the resonance of a number of strings: "all the strings and nothing but the strings" that corresponded to the partial tones within the given note. The sensory elements of the ear should therefore behave like piano strings, resonating specifically for each partial tone they detected. The whole ear had thus become a sort of Fourier analyzer. But was the ear built to do what the theoretician expected of it? What was known about the ear in the mid-nineteenth century? Helmholtz had access to the remarkable research paper on the organ of hearing in mammals that had just been published, in 1851, by Alfonso Corti,[28] and he used the data in his *Theory of Music*. The anatomical results were, naturally, followed by other more complete studies, and Helmholtz incorporated them in later revisions of his book; even the sixth edition of 1913, revised by his students, followed the tradition. We restrict ourselves here to data from the first edition of 1862 because their relatively rudimentary character best illustrates the extraordinary coherence of his scientific research.

The organ of hearing is the *cochlea* in the inner ear (figure 10.7). It consists of a membranous tube filled with liquid, and it forms a spiral like a snail shell. It is enclosed in a curious bony cavity of the same spiral shape deep in the temporal bone. The cochlea is divided into three canals. The first two, the *scala vestibuli* and *scala tympani*, communicate with each other at the apex of the cochlea. At their lower ends, they are separated from the tympanic cavity by the *oval* and *round windows*, each closed by a flexible, elastic membrane. The third, the *cochlear duct*, contains the actual auditory receptors. The air-filled tympanic cavity, in the middle ear, is intermediate between external sound sources and the cochlea, between the *tympanic membrane* (ear drum) on the outside and the oval and round windows on the inside. The cavity contains a series

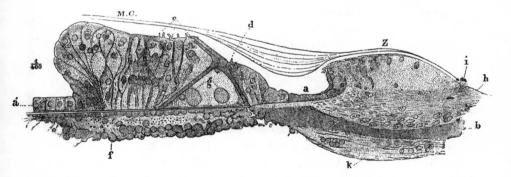

Figure 10.7
A vertical section of part of the cochlea of the inner ear showing the organ of Corti in a schematic drawing from Helmholtz's "Theory of Music." On the left is the outer zone of the cochlea and on the right is the central axis. The tunnel of Corti is visible at the center of the figure (g) between the inner or ascending and outer or descending pillars, which Helmholtz considered a sort of violin bridge able to resonate with the vibrations of the basilar membrane (the lower membrane on which the internal structures are placed). To the right and left of the tunnel of Corti can be seen the hair cells, respectively, inner and outer (of which the latter are much more numerous than the former), which, as we now know, are the real auditory receptors. In the right lower part are several nerve fibers (b). The upper structure marked M.C. (membrane of Corti) is now referred to as the tectorial membrane and is known to be in contact with the ciliated tips of the outer hair cells. (Helmholtz 1863)

of three ossicles—the *malleus, incus,* and *stapes*—which transmit the vibrations of the tympanic membrane to the oval window. So the stapes transmits the vibrations of the air to the liquid in the scala vestibuli with the same frequency and almost the same waveform. Because liquids are incompressible, the vibrations run all the way along the scala vestibuli to the apex of the cochlea, after which they run back along the scala tympani to be absorbed at the round window, which can expand into the air-filled tympanic cavity. Thus, sound stimuli cause vibrations in the liquid of the cochlea, which are then analyzed by the receptor organs in the cochlear duct.

Helmholtz took up Corti's original description of the cochlear duct. Two parallel membranes are stretched between a bony crest arising from the central bony axis of the cochlea and the bone on the opposite side of the duct. The lower *basilar membrane* is taught but elastic and separates the cochlear duct from the scala tympani. The upper *vestibular membrane* (of Reissner) is very fine, except at its base, where it forms a swelling. For the anatomist Reissner, this membrane did not constitute the limit between the cochlear canal and the vestibular membrane, but rather a third membrane with the same origin as the membrane of Corti at the level of the central bony axis, but that was attached much higher than the latter on the exterior wall of the cochlea.

Between the basilar and vestibular membranes, there are various cells mixed with numerous nerve fibers running to the central bony axis. But a rather extraordinary structure immediately attracted Helmholtz's attention, the so-called *arches of Corti*. They consist of two rows of fiber-like cells, the *pillar* or *rod cells*. One row forms the *inner*, or *ascending*, pillar cells, and the other forms the *outer*, or *descending*, pillar cells. Inner and outer pillars bend to touch each other at their apex. These rows of several thousand cells extend side by side from the base to the apex of the cochlea, forming the triangular section, spiral *tunnel of Corti* between them. For Helmholtz, the *organ of Corti* was obviously suited to receive the vibrations of the basilar membrane and then vibrate itself. The most likely theory for him was that the inner cells formed a sort of elastic frame, whereas the outer cells, thin and flexible like cords, connected with the center of the basilar membrane, vibrating when their extremities were shaken by movements of the membrane in response to sound stimuli.

Thus, Helmholtz put his finger on a structure that had all the necessary pieces to act as the resonator, which he needed for his theory of audition. He acted with the authority and conviction with which we are now familiar, although in all honesty his only tool was the microscope, and contemporary science still did not help understand how the vibrations were achieved and what the degrees of tension and flexibility of these supposed resonators were. Further, he still had to demonstrate that every tone had its corresponding specific resonator, making a sort of sensory keyboard out of the succession of arches of Corti. Despite the impossibility of an experimental proof of the resonator theory, he showed no lack of resource in terms of observation and logic. For instance, he had noticed that when a trill was made up of a succession of two musical notes at a frequency of ten per second, it was possible to distinguish the two notes as long as they were at the medium or high frequencies of the auditory spectrum. In contrast, when the notes were low, the trill became progressively more blurred, the notes were confused, and they became harsh and unpleasant. Helmholtz attributed this phenomenon to general properties of resonance of elastic bodies, with the trills becoming more blurred when the duration of residual resonance became longer and remaining distinct when the latter did not exceed that of the notes sounded. Of course he saw therein the proof that there must be different structures in the ear that vibrated to sounds of different frequencies.

But there was yet another possible difficulty: William Thierry Preyer[29] had shown that it was possible to discriminate a change in pitch of half a cycle at around 1,000 cycles per second. According to Heinrich Wilhelm Waldeyer,[30] there were 4,200 Corti fibers in the cochlea, which left 600 fibers per octave or 50 fibers per semitone. Even if that was too few to imagine that there would be one fiber corresponding to 1,000 cycles and another to 1000.5, it did not matter, Helmholtz said, because a sound that fell between two neighboring fibers vibrated both fibers at the same time, but the one that was closer to the sound vibrated more. So discrimination of small differences in pitch in the interval between two fibers depended on the finesse with which one could compare the intensity of excitation produced by the two corresponding nerve fibers. In this way, one could explain that, as sound increased in frequency continuously, our sensation was equally modified continuously, not by sudden jumps as would happen if there were only ever one Corti fiber vibrating at a time. One can only admire the scientist's elegant intellectual gymnastics, once again calling on psychological arguments to support his physiological theories, which he was absolutely convinced were correct.[31]

In later editions of Helmholtz's book, several elements of major importance were added to those of the first edition. First, the observation of Victor Hensen[32] and Carl Hasse[33,34] that the basilar membrane increased in width from the oval window to the apex of the cochlea, accompanied by a progressive increase in the size of the arches of Corti, doubtless constituted a further important argument in favor of selective resonance in different parts of the basilar membrane. Second, Hasse emphasized the possible role in audition of tiny "hairs" (in fact, cilia) on what are known as *inner* and *outer hair cells*, situated just alongside the inner and outer pillars. These cells were richly innervated and covered by the *tectorial membrane* (of Corti). Vibrations of the hairs might be of great importance in the genesis of nerve impulses. At first Helmholtz underestimated the possible role of these hairs. However, when he discovered that the cochlea of birds did not contain arches of Corti but did have numerous hair cells, he finished by admitting that they might be more important than the arches in the transformation of resonance to the sensation of sound. Further, after the first edition of Helmholtz's book, Hensen[35] discovered in the Mysis shrimp tiny hairs on the free surface of the body, arranged from large to small and thick to thin, that he was able to demonstrate as playing a role in hearing. In particular, Hensen succeeded in transmitting the sound of a valve horn through water in a trough to the shrimp and observed with a microscope that a given tone caused certain hairs to vibrate, whereas other tones affected different hairs.

Helmholtz, in tackling audition, behaved in just the same way as previously when he produced his study *On the Conservation of Force*: profound theoretical conviction combined with exceptional competence in physics, mathematics, and laboratory technology. Just as for vision, he was in no way as well equipped for experimentation as we are now, and he had to be satisfied with what was known in his day. But he did not hesitate to go beyond what strict empiricism might allow him to state by basing himself solidly on the theoretical model that he had assimilated.

The Perception of Timbre

To use his own terminology as set out in the introduction to his book, sensation referred to a physiological level and perception a psychological level. He called sensations the impressions produced on the senses in as much as they appeared as states of the body; on the contrary, he called them perceptions when they served to provide representations of external objects. In this sense, the sound of a violin—the fundamental accompanied by its harmonics—provoked many sensations as there were specific resonators in the inner ear vibrating to each partial tone. On the contrary, fusion of the harmonics into a single sound, thus enabling it to be recognized as that of a violin, made a perception. He recalled that each partial tone existed in the complex sound of a violin, just as the various colors of the rainbow existed in white light. Just like the fundamental colors of white light, partial tones were difficult to hear, even for accomplished musicians, unless using suitable apparatus such as resonators, as well as their undivided attention. It was fortunately not indispensable to perceive partial tones because the essential was to recognize the sound of a particular musical instrument. Even if much information was thus lost in the passage from sensation to perception, it was not important because perception was a concept that essentially fulfilled the criterion of usefulness.

In addition, one recognized the sound of a musical instrument, such as a violin, because one remembered, from having heard one on many occasions, that a particular association of partial tones could only come from a violin. This therefore supposed learning by association based on practice, comparison, and habit. Recognizing individually several musical instruments playing together, or listening to a speaker in a noisy meeting, involved gathering scattered sensations into coherent perceptions and paying attention preferentially to one of the sound sources to the exclusion of the others.

We may note how Helmholtz considered that the transition from sensations to auditory perception involved, as for vision, unconscious processing of the information, with consciousness only entering the scene later. We may also note his insistence that this processing implied only acquired factors to the exclusion of possible innate ones. Nevertheless, he was obliged to admit that certain factors in this unconscious processing were of a structural nature. For example, he noticed that the pitch of a complex sound was always determined by its fundamental and not its harmonics. Another example was that the overall intensity of a complex sound was always equal to the sum of the intensities of its partial tones.

In Conclusion: No Bells in the Orchestra

At this stage of our account, we must agree that Helmholtz achieved a remarkable work of synthesis, from acoustics to the physiology, and even the psychology, of auditory perception. But he had not yet ventured into the domain of musical esthetics or risked himself in that of beauty and its possible relationships with the laws of the natural order. Certainly, he had greatly demystified the problem of Pythagoras and the mysterious, if not mystical, interrelationships of consonant sounds. Thanks to his physical study of complex sounds, he confirmed their wave-like nature and showed that they were composed of partial tones or harmonics of which the frequencies were always a multiple of the corresponding fundamental tone. So the mysterious problem of Pythagoras was reduced, at least partly, to a sequence of exclusively physical factors associated with the presence of harmonics. However, Helmholtz had not yet tackled the problem of consonance as such and its relationship to Pythagorean ratios.

One could easily admit that a fundamental note played with its octave (C0-C1), its fifth (C0-G1), or its third (C0-E2) would be consonant: Indeed, did one not call the chord CEG a perfect chord? However, it was still difficult to understand what made them consonant, and therefore pleasant to hear. Were they pleasant because we were used to hearing them together of necessity in complex sounds? Or were they consonant simply because they were not dissonant and why? Helmholtz was obliged to study consonance and dissonance, and paradoxically it was thanks to the latter, as we discuss in the next chapter, that he was able to find a certain scientific basis for the concept of consonance.

Before explaining the results of his research on dissonance, he invited his reader to a conducted tour through the orchestra, studying for each instrument the composition of its sound in terms of harmonics. It was a truly extraordinary tour not only because it was the first quantitative study of its kind, but also because of the amount of detailed work necessary to accomplish the task. Here are a few of his observations:

Simple sounds, like those of the great flue pipes of an organ, were soft, gracious, and not in the least harsh. The music of a flute was close to a simple sound.

Sounds accompanied by lower harmonics, up to the sixth, and of medium intensity were full and rich, soft and harmonious. These were the sounds of the piano, the open pipes of an organ, the soft, quiet sounds of the human voice, and the horn.

The sound of the clarinet owed its hollow and somewhat nasal character to the absence of any even harmonics. When harmonics above the sixth were noticeable the sound became sharp, strident, and harsh, whereas at a low intensity these harmonics increased the character and expressivity of the music, as was true for bowed instruments, reed instruments such as the oboe and bassoon, and the human voice. In this category, too, were also brass instruments, which, played at a high intensity, could produce harsh and penetrating sounds.

He produced a truly exemplary study of the violin thanks especially to the construction of a *vibration microscope* (figure 10.8), which enabled

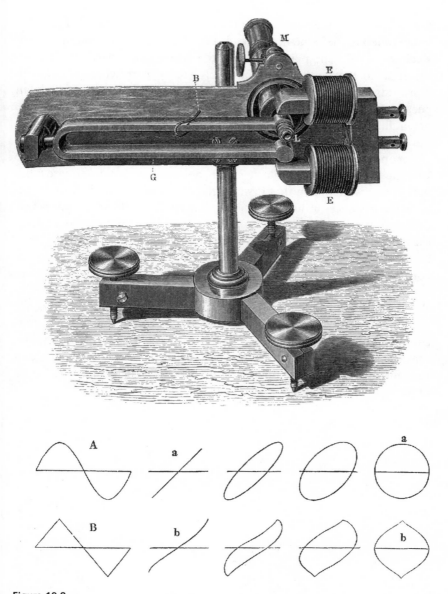

Figure 10.8
The vibration microscope used by Helmholtz in his studies on musical acoustics to analyze the composition of sounds produced by string instruments, especially the violin. The vibrating cord was visualized though a microscope that was also oscillating. By persistence of images in the observer's retina, variable figures were produced as a consequence of the combination of the two movements. The shape of the figures depended on the frequency, amplitude, and relative phases of the two oscillations.
(Helmholtz 1863)

him to compare the vibrations of a tuning fork with those of a violin string tuned to the same note. He observed that the first harmonics were weaker than for the piano, but that the higher ones, particularly from the sixth to the tenth, were noticeably stronger, which explained the typical stridence of bowed instruments. He also studied the bow in detail because its role was not only to produce vibrations in the strings but also, depending on the musician's talent and a good elasticity of the wood of the violin, to ensure the regularity of the sound without any scraping.

Among the instruments he studied, we should mention one because what he says about it is quite revealing about his esthetic standards: the bell. It was well known that it was difficult to cast a good bell to obtain an equal thickness around its whole circumference. If the thickness was different at two different places, there would be a spot on the edge of the bell that would vibrate to give a certain tone while another neighboring spot produced a slightly different tone, whereas the intermediate zone between the two produced both tones at the same time, "causing the beats that are heard from most bells as their sound gently subsides."[36] We should note here that these *beats* were for him the principal cause of unpleasant dissonance, as we see in the next chapter. When one considered that bells usually had a fundamental mixed with harmonics that were not related to it, the result was that when the harmonics and fundamental were close to each other, the resulting sound was highly unmusical and unpleasant to the ear.

11 The Musical Ear

... that music is a language by means of which messages are elaborated of which some at least are understood by the immense majority of people whereas only a tiny minority is capable of producing them, and that of all languages this one alone unites the contradictory characteristics of being at the same time intelligible and untranslatable, makes of the creator of music a being who is the equal of the gods, and of music itself the supreme mystery of the sciences of mankind, the one against which they stumble and which contains the key to their progress.
—Claude Lévi-Strauss[1]

Why Music?

Music is as difficult to define as it is indispensable to man. It is yet another great mystery for the biologist to understand why, from being completely absent in nonhuman primates, it appeared in *Homo sapiens* under the influence of evolutionary pressure comparable to that which gave rise to the blossoming of language and consciousness in our remote ancestors. Furthermore, music, as Helmholtz knew and enjoyed it in the1860s, has continued to evolve to the present day, with dismantling of harmonic structures, challenges to melody, and the invention of serial or atonal music and complex or random rhythms. However, one thing has scarcely evolved since the appearance of the phenomenon of music: the inevitable emotional shock that it impresses on our conscious and unconscious mind, a sort of waking dream, with immediate consequences for our psyche, such as feelings of ease or distress, exaltation or depression, pleasure or displeasure. In Helmholtz's day, when the so-called classic rules of musical composition in Europe were almost unanimously accepted, the classification of chords as consonant or dissonant was of the greatest practical importance. Helmholtz's great merit was to have undertaken theoretical and experimental research in an attempt to show

how physics and the ear working together determined the consonant or dissonant character of chords. But we discover, as we examine Helmholtz's results, his dilemma in being unable to dissociate his view of aesthetics, which strived to respect the genius and creativity of the artist, from another view, steeped in nostalgia for a mechanistic approach to musical composition, which in his opinion would have better respected the laws of physics.

Consonance and Dissonance

The definition that Helmholtz gave at the conclusion of his research was brief and to the point: "Consonance is a continuous, and dissonance an intermittent sensation of sound." This definition, based on experimentation, was basically similar to that of Euclid: "Consonance is the mixing of two tones, one high, the other low. Dissonance on the contrary is the impossibility for two tones to mix without producing a harsh impression on the ear."[2] Helmholtz had the audacity to explain this distinction between consonance and dissonance by the physical phenomenon of *beating* when two tones of almost identical pitch are played at the same time.

Let us look at this more closely. Two tones of similar pitch, which therefore stimulate the same auditory fibers, do not necessarily produce the same sum of sensation that each would produce separately. As physics shows, the intensities of two tones in temporal phase summate, but if the two tones are completely out of phase, their intensities cancel each other and you hear nothing at all because of the classic phenomenon of interference. Two tones of slightly different pitch, however, produce interference of a more complex type that we call *beats*. We can represent this from the following drawing based on Helmholtz[3]:

A—B = 18 waves

A—B = 20 waves

During the time A—B, eighteen waves occur above the abscissa and twenty waves below. It is not difficult to imagine from the principle of interference that the intensity of the sound is doubled at points 1, 3, and 5 because the waves are in phase, whereas at points 2 and 4 they are completely out of phase, and so no sound is perceptible. We call beats the succession of loud sounds heard at points 1, 3, and 5 and the moments of relative quiet between them. The number of beats in a given time is thus equal to the difference between the respective number of vibrations during the same time by the two sounds, that is 20–18 = 2 (time zero obviously not being taken into consideration for this calculation).

Based on this theoretical concept, Helmholtz constructed a double siren consisting of two Dove sirens[4] with several series of holes in each disk. They rotated at the same speed, but one could manually rotate one of the two air chambers injecting air through one of the disks such that it produced a sound of a slightly different frequency. He was thus able to confirm the rule *n + x vibrations per second—x vibrations per second = n beats per second* and so test all possible combinations of chords.

To summarize the essentials of his observations: When the frequencies of two tones were so close that the beats were only three or four per second, the effect could be pleasant because it resembled the warm-sounding tremolo of a bowed instrument. There even existed an organ register that relied on the vibration of two pipes of similar pitch, which gave a trembling sound thanks to this same beating phenomenon. On the contrary, when the difference in pitch increased, the beat frequency increased, until when it reached 20 to 30 per second, the sound became confused, harsh, and penetrated by very unpleasant rumbles. The rumbling consisted of intermittent bursts of sound, rather like the letter "R." It was worst at a beat frequency of 30 to 40 per second. If the frequency increased even more, the sound became less harsh and more continuous, and the beats finally became inaudible.

In practice, it was a little more complicated: The harshness of the beat depended on both the dimension of the sound interval and the frequency of the beats. If the beat frequency were constant (e.g., 33 beats per second), the harshness increased by the fifth C0-G0 up to the semitone B2-C3, passing though all the intermediate intervals. This explained why it was scarcely possible in harmony to utilize half-tone intervals in the upper octaves. It also explained why the lower an interval was, the greater it had to be to hope to remain consonant. Helmholtz noted that upper thirds were pure but the lower thirds harsh.[5]

Combination Tones

As we have just seen, for Helmholtz, consonant and dissonant chords differed essentially in that the former were perceived as continuous sounds, whereas the latter merged into a vague mass of sound interrupted by pulsations or beats at an unpleasant frequency. The ear was unable to perceive separately the two original sounds. But it was a completely different story for another sort of simultaneous sound called a *combination tone*.[6] Two tones that were clearly perceived as different in pitch gave rise to a third, perceived separately and quite distinct from the other two. For this third tone, the combination tone, to be audible, the two first ones must be played forcefully and for a prolonged period. Combination tones were discovered in 1740 by Georg Andreas Sorge, a German organist, and were the subject of an important study by Italian violinist Giuseppe Tartini,[7] who was, incidentally, the composer of the well-known *Devil's Trill* violin sonata. These two only described a single type of combination tone that Helmholtz called *a difference tone*. He, however, discovered others, more difficult to hear, which he termed *sum tones*.

How did he produce these two types of combination tones? He used two prolonged loud tones with a harmonic interval of less than an octave. He first played the lower tone and then added the higher; by listening carefully at the moment the high note sounded, he heard a weaker, lower note that was precisely the combination tone he was seeking. Difference tones had a frequency equal to the difference between the respective frequencies of the two primary tones. For example, for the fifth C1-G1, the difference was 396 (G1) − 264 (C1) = 132 (C0). As to sum tones, which were much more difficult to detect, their frequency was equal to the sum of the frequencies of the two primary tones.

What is quite remarkable is that sum tones had remained completely unknown until their discovery by Helmholtz. He postulated their existence based on mathematical and physical models, but they were so difficult to perceive that he only heard them later, thanks to his polyphonic siren. Is it not like the way in which, almost a century later, the existence of the planet Pluto was proposed thanks to mathematical models, only to be confirmed later by telescopy?

But there is another point to emphasize: At the time of Helmholtz, many people thought that combination tones were purely subjective and originated in the ear. But he demonstrated that difference and sum tones really existed objectively, outside the ear, first because the mathematical model provided formal proof of their existence and then because he had succeeded with his double siren and suitable resonators to isolate the two primary sounds as well as the combination tones. The only condition necessary for the latter to appear was that the same volume of air had to be vibrated by the two primary tones, as was the case for the poly-phonic siren in which the same disk contained several series of holes that could receive the air from the same chamber at the same time. Another example was the harmonium, which depended on a single source of air and from which combination tones could readily be heard.

Finally, we may recall that Tartini advised violinists to always tune their instrument by listening carefully to the combination tones coming from their single sound box as they bowed two open strings at the same time.

Diabolus in Musica

Experience tells us that beats not only occur in chords composed of simple fundamental notes, as in the examples so far, but that all harmon-ics and even combination notes can cause beats. We can easily imagine Helmholtz's enormous work in modeling mathematically all imaginable cases of interaction and testing them with real cases of classical harmony as experienced by musicians. The agreement between his theory and practice was remarkable in every way. For example, the ideal consonant chord, in theory as in practice, was when all the harmonics coincided, as was the case of the *unison*, where two fundamental notes and their har-monics were identical. The octave (C0-C1) was almost perfect as was the twelfth (C0-G1), where the higher note coincided with the second har-monic of the lower note. Within an octave, the fifth had the best conso-nance (C1-G1) because the dissonant harmonics of these two notes were too high to be really annoying. Then came, in order, the fourth (C1-F1), the sixth (C1-A1), and, finally, the third (C1-E1). The upper thirds were pure, but the low thirds were harsh. Indeed, until the end of the Middle Ages, the third was considered dissonant because, sung by male voices, it was of necessity rather hard.[8] A well-known example of a third is the beginning of Beethoven's Fifth Symphony: *po po po pom, po po po pom*. The first *po po po pom* contains an interval of a major third between the first three and the fourth note, and the second *po po po pom* is an interval of a minor third.

The worst dissonant tones were the minor second (C1-D flat 1) and the infamous major seventh (C1-B1); in both there was only a semitone between the two notes of the chord, either between C1 and D flat 1 or between B1 and C2, which was the harmonic octave of C1 and was heard at the same time. So Helmholtz was not far away from the opinion of ancient authors such as Franco von Köln, who in the twelfth century considered thirds as imperfectly consonant and sixths as absolutely dissonant. As such, he agreed with the Swiss musician Henricus Glareanus who, in his *Dodekachordon* of 1547, classified consonance in a similar fashion on harmonic criteria, whereas Helmholtz classified consonance essentially on mathematical, physical, and experimental criteria. Furthermore, he paid homage to Rameau and d'Alembert, who, for the first time, based their studies of consonance on scientific rather than metaphysical criteria. He said that his own concepts were "based entirely on a careful analysis of auditory sensation, but even if any practiced ear could have achieved the same without the help of theory, the guidance of theory and the help of appropriate methods of observation made it extraordinarily much easier." He continued, "I do not hesitate to state that the research I present reveals true and sufficient causes of consonance and dissonance of musical sounds, based on a precise analysis of tone sensation, and on purely scientific, rather than esthetic, principles." It was perhaps a little hasty as a conclusion, but we look at it again later.[9]

But what was the role of the ear in the perception of beats? On this topic Helmholtz was quite definite: Here was an exception to the law according to which two sounds or two auditory sensations could coexist without acting reciprocally on each other; if the Corti fibers were in principle stimulated by both sounds separately, in this case, they mutually reinforced or weakened each other so that they produced a single series of impulses in the nerve. What made beats unpleasant and thus caused the sensation of dissonance was their intermittent character, "like flickering light or scratching for the skin," which did not allow the sense organ time enough to adapt or habituate.[10]

Consonance was only defined by the fact that it was not dissonant. We can see how difficult it was for Helmholtz, who only disposed of the conceptual tools of his day, to define such an essential characteristic of our mental activity as pleasure. Maybe it is nothing more than the absence of displeasure. Helmholtz wrote that, for Euler, "The mind would perceive as such the rational relationships of sound vibrations" and derive pleasure from that. On the contrary, for Helmholtz, "it only perceives a physical effect of those relationships, the intermittent or continuous sensation of the auditory nerves."[11] The ear almost had the status of the brain for it *knew* the consonant or dissonant character of musical chords.

What, then, should we think of the *diabolus in musica*, that awful tritone that has earned the scorn of generations of musicians? A *tritone* is an interval of three whole tones, such as F1-B1 (FGAB), which is really an augmented fourth, classed by Helmholtz, and also much later by Paul Hindemith,[12] as one of the most dissonant chords in existence. Furthermore, B1 is only a semitone from C2, an octave above C1 with which it makes a harsh and dissonant seventh. But its legendary harshness and Helmholtz's scientific aesthetic principles have not prevented the tritone from becoming a favorite chord of impressionist musicians and its featuring as a fashionable devil in Claude Debussy's *Pelléas et Mélisande*.

The Search for the Foundations of Modern Music

At last it is time for us to talk about music. Until now, we have been limited to purely scientific discussions: the wave nature of sound, timbre as defined by harmonic spectra of a fundamental note, the anatomy and physiology of the ear, perception stemming from elementary sensations generated by the cochlea, and the physical and physiological basis of consonant and dissonant chords. But as Helmholtz said, "If in the theory of consonance we have spoken so far of pleasant and unpleasant, this was only a matter of the direct sensory impression produced on the ear by an isolated harmony, without any reference to artistic contrast or expression, which means only about the comfort of the senses and not about esthetic beauty. These two must be carefully distinguished, although the former is an important means of achieving the goals of the latter."[13]

The degree of harshness of a sound depended on the anatomical structure of the ear and the presence or absence of beats. It was an objective fact. Nevertheless, continued Helmholtz, there was variation in the taste of individuals or nations that allowed one to understand why, in experiencing music, a listener accepted to tolerate this harshness to a greater or lesser extent and why the limits of consonance and dissonance varied according to habits or culture. The "system of scales, of keys and their harmonic structure does not only rely on invariable natural laws, but on the contrary, at least in part, it is the consequence of esthetic principles that have given rise to change with the progressive development of humanity and will continue to do so." One cannot help but agree with these conclusions, but we see later that for their author these aesthetic principles were subject to certain paradoxical limitations.

Before coming to that, let us follow Helmholtz in his overview of twenty centuries of musical history that he summed up in three successive musical periods: *homophonic*, *polyphonic*, and *harmonic*. There is little doubt that, since he published his great work, substantial progress has been made in this domain, but his text is still fascinating to read because of his coherent and logical sequence of discoveries in terms of musical expression. The author guides us through the meanders of comparative musicology, exposing his own basic principle for the development of European music "that the whole mass of tones and harmonic liaisons must bear a close and clear relationship to a freely chosen tonic note, from which the composition of the whole musical phrase must develop and to which it must return." He added, "But as we can see, this principle is esthetic and not natural."[14]

The Invention of the Tonic Note in Homophonic Music

Terpander, born in 676 BCE in Antissa in Lesbos a century before Pythagoras (560 BCE), was the first known Greek musician. At that time, there were already folkloric and, especially, religious songs. It is said that Terpander invented the seven-cord lyre. In addition, he seems to have been an innovator in composition by deciding a fixed structure for the *nome*, a sort of vocal solo accompanied by a cithara. Epic-style texts were preceded by a short, austere, and unmodulated instrumental prelude. More varied, free musical styles only developed later. For example, in the *dithyramb*, also originally codified by Terpander, in which the cithara accompanied the choir during religious or orgiac processions in honor of the god Dionysos, the age-old habit of a dialog in sung verse between a director and a chorus resulted for Aristotle in the emergence of the tragedy.

What was remarkable in Greek music, said Helmholtz, was that it should have remained essentially homophonic—that is, without chords. With the exception of short pieces played as solos, in which a certain liberty and virtuosity gradually developed, the role of the musician was essentially to accompany the verses of a poet in homophonic measures.

To Aristotle's great surprise, musicians only used consonant octaves, to the exclusion of all others that were, however, already known. In fact, Pythagoras had already discovered the mathematical laws underlying the principal consonant chords thanks to the monocord. But his remarkable research was of much more interest to mathematicians and moralists than musicians themselves. Just before his death, Pythagoras exhorted his disciples not to consider the monocord as a musical instrument but as a measuring tool because the harmony of musical proportion matched that of distance between the celestial bodies, knowledge of which brought equilibrium to the soul.

It was only from the fourth century BCE that they began to use consonant fourths and fifths. However, Greek musicians used a large number of different modes in which the octave began and ended with one of the notes of the traditional scale—for example, C0-C1, D0-D1, A0-A1, and so on, which enabled them to give an individual musical atmosphere and character to their compositions depending on which mode they chose. A remarkable fact was that the role of music as a pedagogic and religious tool for the formation of character was such in Plato's day that he did not hesitate to reject Ionian and Lydian modes, which produced effeminate music, and instead only accepting Dorian and Phyrygian modes that, respectively, instilled courage and sobriety. Even the reasoning mind of Aristotle appealed to an ethical sense in matters musical and preferred above all the Dorian mode.

In summary, Greek music was above all a servant of the text, without much intrinsic value, without harmony, but expressing an individual character thanks to different modes.

The question that Helmholtz asked next was whether there was in this music "a particular relationship between each note of the scale and a single fundamental: the tonic note."[15] Indeed, in the modern music of the 1860s, the tonic note represented the link between all the notes of a phrase. For François-Joseph Fétis, the Belgian musician and musicologist,[16] it was in nations where tonality based on a tonic note as the cohesive principle had dominated that harmony and melody had developed.

As we saw earlier, the tonic note is the first one of the major or minor scale. In classic harmony, all melody began and ended with the tonic. To the nonmusician who might be frightened by this jargon, I suggest singing to yourself, for example, the well-known air "Frère Jacques," which begins and ends with the same note, the tonic of C major, and you should immediately see intuitively the importance of this rule for the cohesion of the melody.

As the alert reader will have certainly noticed, Helmholtz was seeking in Greek music the slightest evidence that would enable him to understand the evolution of music toward modern harmony, of which the rules were of capital importance to corroborate his physiological theory of music. Some of this evidence was even provided by Aristotle, for whom the *mese* (from μεση, the keynote, Helmholtz's *Mittelton*, and equivalent to the modern tonic note for him) was the most important note of the melody. Helmholtz cited Aristotle: "All good composers use the mese often, and if they abandon it they soon return, more than for any other. ...

The mese is a link for sounds, especially beautiful ones." Aristotle "compared the mese with a conjunction in language, especially those that signify 'and', without which language could not exist. ... Pythagoreans compared the mese to the sun, and other sounds of the scale to the planets."[17]

Songs always began with the mese or tonic. However, unlike our own custom, they ended with the fifth lower than the tonic (e.g., G0 below the tonic C1). The musical phrase was subject to the normal inflexions at the ends of phrases and thus resembled a Gregorian chant.

The Slow Blossoming of Polyphony and Harmony

It was only in the ninth century that harmony was formally born by the bold hand of the monk Hucbald of the Abbey of St. Amand in Tournai. Hucbald was the first to describe, under the name of *diaphonia* or *organum*, a new process of composition in which a sacred phrase was accompanied in parallel by a second voice singing the same phrase either at the octave, fifth, or fourth. Thirds and fourths were still considered dissonant. Unfortunately, the composer only rarely changed the interval during the same melody, which for our modern ear is terribly monotonous.

A little later *descant* was born, another type of polyphonic music in which two completely different airs alternated, with small modifications of rhythm and pitch, but ended together in consonance. In many cases, the two voices sang together but in opposite directions, the first rising while the other descended and vice versa. Helmholtz explained that this was less to achieve consonance than to avoid dissonance.

Organum and descant underwent considerable developments: We might mention here the French composer Pérotin (ca 1230) at Notre Dame in Paris, who wrote several liturgies for two, three, and even four voices, ornamented with new rhythmic figures. His *Alleluia Nativitas* was one of the first masterpieces of Western polyphony.

The development of such new rhythms at this time resulted notably in the spectacular discovery of the *canon*, in which different voices were introduced one after the other, each rigorously taking up the same musical phrase of the start, whether at the same place in the scale or not. For the first time, Helmholtz continued, music freed itself from its total servitude to the text that homophony imposed and thus enriched itself with a structure for authentic musical expression. Indeed, in the sixteenth century, madrigals and motets using canons, and later fugues accompanied by instruments abounded.

Decisive steps in the development of real harmony were the elaboration by Jacopo Peri,[18] probably born in Rome but working as organist
and singer in Florence, of the recitative accompanied by simple chords,
and also that by Lodovico da Viadana[19] of the *basso continuo* played by
the organ or harpsichord weaving long consonant chords into the harmonic substance of the song. At least this was Helmholtz's opinion, but
it is disputed today. The *basso continuo* was the lowest pitched part of a
piece of music, of which the principal function was to support the overall
harmony.

This slow blossoming of harmony—thanks to the development of
polyphony, of rhythmic structures such as the canon, and the conception
of the continuo—was, however, still a mere hesitant stutter. Despite the
real masterpieces produced during this period, the development of
harmony remained effectively blocked, the cause being manifestly the
chaos that reigned in the use of modes inherited from Greece and the
Gregorian Middle Ages. They were so numerous that different modes
were sometimes used for each voice of the same polyphonic chant, so
much so that it was impossible to say in which principal mode it had been
written. It is obvious that in such conditions it became difficult to detect
Helmholtz's cherished tonic tone even within Aristotle's limited meaning.
Reform of modes was therefore necessary; it took place slowly and
resulted, thanks to Glareanus (*Dodekachordon* of 1547) and Gioseffo
Zarlino,[20] in the disappearance of most traditional modes and their
reduction to our modern keys and major and minor scales, formerly
Ionian and Aeolian modes. Now let us see how harmony developed
conjointly with the tonic note and melody once the major and minor
scales had been accepted.

The Flourishing of Harmony: Martin Luther and Claudio Monteverdi; Palestrina Censored by Helmholtz

Who would have believed that, in introducing religious reformation,
Martin Luther[21] would equally have become the father of a great musical
reform for which melomaniacs of subsequent centuries would be unanimously grateful? What was it about? We must remember that Luther
translated the Latin Bible—the fundament of the true faith—into the
vernacular, and that he firmly desired that members of the reformed
church should forthwith have an active role in the liturgy, singing sacred
texts in their own language. The fashionable polyphony of the time was
excluded, being corrupted by its traditional links with the Catholic
Church. It is fortunate that Luther liked music and that he had taste.

Music was a fair and lovely gift of God,[22] he said, because it drove away the devil and made people happy. Composer of the first German Mass in 1526, he inspired a vast movement of composition of sacred songs, chorales, and motets, in addition to those he wrote himself. The Protestant chorale had a simple harmony, and all the voices sang in unison. Further, the fact that sacred songs usually adopted well-known popular airs of resolute tonality was really revolutionary, but it ensured that their pious nature was all the more appreciated. The cohesion of the chorale, without flourishes or complex rhythmic structures, was for once ensured by an honest harmony based solidly on the tonic note. At the same time in France, the Calvinist Clément Marot[23] produced his famous translation of the Psalms in a similar spirit.

In contrast, in the Catholic world, the situation was less straightforward. At the end of the Council of Trent in 1563, Pope Pius IV had ordered Giovanni da Palestrina to reform religious song. A former chorister at the Sistine Chapel, he was, however, more used to composing for accomplished chorales than for large lay congregations. Thus, one can understand that in his vocation he concentrated above all on the patrimony of Gregorian plainchant and polyphonic music. For Helmholtz, this had not allowed him to develop harmony as well as his reformed colleagues, and he did not hesitate to reproach him vigorously for it. For example, concerning the first strophe of his *Stabat Mater*, he administered Palestrina a real lesson in composition: "But it is only at the eighth bar of the phrase that a chord in D minor appears, that a modern composer would have placed at the first suitable moment in the first bar … what we regret is firstly that the tonic chord does not play from the beginning of the phrase the primordial role that it is given in modern music." Helmholtz indeed had his knife into anyone who did not follow his way of thinking. Fortunately, he told us, from the seventeenth century on, the development of opera with the choir singing polyphonically or solo voices accompanied by a continuo gave a considerable impulse to harmonic music. For him, "in Monteverdi, who was extraordinarily fertile in new ideas, we find the first free use of a chord of seventh" and "there quickly developed an audacious use of dissonance … in order to better accentuate the nuances of expression and not as chance accidents."[24]

Melody and Mechanics

This being said, the irresistible rise of harmony was not only that of the indispensable principle of vertical organization, in which each note of a song was situated in the heart of a suitable chord, but remained independent of those around it. Horizontal cohesion, that is, successive chords between themselves, was just as necessary to harmony: The tonic note and its consonant chord contributed effectively by limiting the musical phrase at its beginning and its end. But there was, above all, the melody that was both the musical phrase and structural basis of its own cohesion. Melody deserves our careful consideration. For Helmholtz, it was defined as a "movement" of notes whose pitch varied constantly; he noted in this context that in the music of all known peoples, the pitch of the notes in melodies varied in steps and not continuously. The melody directly mirrored the state of mind of the composer. For this state to become the subject of an immediate sensation for the listener, it was necessary for the duration of each note as well as the regular succession of strong and weak accents in the melody to be measured, imposing tempo to a piece of music. Helmholtz referred directly to Aristotle, who defined the movements of rhythm and melody as "energetic."[25] Furthermore, the variation in pitch of the sound had to be in measurable steps because, just as the movement of the music could only be quantified if its sequence was split into divisions, the speed of a movement was measured by the ratio of the length of the music to the time taken. Yes indeed, the melody was seen as a movement with a certain speed, and obviously at the source of this movement was a certain force. Helmholtz spoke of this more than once, but we have to admire the extraordinary and uncompromising technician in him as he struggled with the significance of musical expression.

Beyond these mechanistic allegories, we have just seen an important point: Melody depended on the impulse of an interior force and from that derived all its power to give cohesion to a musical phrase. However, this cohesion could only manifest itself by the quality of the links between successive chords of the melody, from the first to the last tonic note. Therefore, the whole science of harmony—and Helmholtz devoted an exhaustive technical description to it—consisted of ensuring the harmonious passage from one chord to another, bearing in mind that all chords consisted of several fundamental notes, each with harmonics and combination tones, and that the memory of a chord was still present when one played the following chord.

In a melody, the notes and their chords were like lime trees along the drive of a country house: Each tree had its intrinsic beauty, but their disposition along the drive also had to be harmonious and balanced. For us, however, it is clear that music must respect the principles of harmony, and thus the scientific rules that distinguish consonance and dissonance. In this, the ear bears a very great responsibility.

The Scale: From Orpheus' Lyre to Helmholtz's Harmonium

After this brief discussion of the history of music, at the end of which Helmholtz arrived at a sort of apotheosis of modern music supported by a tripod of tonic note, harmony, and melody, it still remained for him to consider a final problem: that of scales. Although these have been an inexhaustible source of controversy since Antiquity, he considered that he had found a definitive solution, consistent with his scientific model of consonance and dissonance, in the natural or "just" scale.

Myth and Truth

The myth of Orpheus was not only that of the fragile triumph of music and love over death: It was perhaps also that of the origins of the scale. At least that was what the Roman poet Boethius related of the Aoidos of Thrace, whose legendary lyre bore four strings, supposedly tuned to C,F,G,C—that is to say, the fundamental with its fourth (F), fifth (G), and its octave (C). As Helmholtz pointed out, nothing is more natural than the octave, just as much for Greek choirs as for modern popular songs. Children spontaneously sang an octave higher than the fundamental sung by men, the whole being perceived as a single sound. After the octave, the fifth and fourth were the most easily accessible: "Even for singers with little experience ... rehearsing the fifth seems natural," and "the fourth is said to have been found in the same way as the fifth, which is the tone of which the first harmonic coincides with the second harmonic of the fundamental. As for the fourth, it is the tone of which the second harmonic coincides with the third harmonic of the fundamental."[26]

The Pentatonic Scale

At first sight, the third (E) and the sixth (A) could have been discovered in the same way, but the common harmonics of the fundamental were weak and difficult to hear even for an experienced musical ear. The third harmonic of the third (E) only coincided with the fourth harmonic of the fundamental C. So, it was necessary to wait several centuries, for polyphonic music, before they were defined precisely.

But then, said Helmholtz, how did the Ancients fill the gaps in Orpheus' scale? Arbitrarily, as in the scale of Ptolemy in the second century and still used by the wandering musicians of Egypt, who divided these musical spaces into equal parts, or in certain Indian scales with their quarter tones? More logical, according to Helmholtz, was the choice of the Chinese and the Gaels who discovered D by increasing C by a fifth and B flat (half a tone lower than B) by increasing F by a fourth. That gave the famous five tone or pentatonic scale (C,D,F,G,B flat,C). Many variations on this exist and are still characteristic of modern Chinese or Scottish music. They are even sometimes used by other western musicians, such as Béla Bartok, Igor Stravinsky, or Antonin Dvorak in his *American Quartet.*

Pythagoras' Scale of Fifths

As to the Ancient Greeks, they were not lacking in imagination. Terpander had already added three extra strings to the four of the classic lyre, tuned to C,D,E,F,G,A,C—that is, the octave minus the seventh B. His initiative was followed by others with more or less success, which served to confirm the arbitrary aspect of the subdivision of the octave or Orpheus' scale.

Then Pythagoras arrived. For him, a ratio of two simply reproduced the same note at different octaves ($2/1 = C1/C0$). It was a ratio of three that was effective in enabling the calculation of the fifth of the fundamental ($3/1 = G1/C0$ and $3/2 = G1/C1$). Number 3 was quasimagical because it was the closest number to 1 that enabled one to find notes other than the octave. Pythagoras' scale was built on this simple model: He calculated a series of successive fifths, each time multiplying the previous one by $3/2$. He obtained C1, G1, D2, A2, E3, B3, F4, which he then transposed to the octave C1-C2. Transposing the fifths in question to the octave C1-C2 meant searching for the equivalent of each one an octave or two lower. For example, to transpose B3 to B1, one had to divide the frequency of B3 by two to obtain B2and then divide that by two to obtain B1.

The ploy succeeded easily because many musical instruments are tuned to successive fifths and are easier to tune correctly than other intervals other than the octave. For example, the violin is tuned to G,D,A,E. It will be obvious to the reader that, in doing so, Pythagoras had neglected harmonic relations other than the fifth, although he had discovered them himself. If he had taken notice of them, his scale would have had different characteristics.

The Scale of Natural Harmonics

In the fourth century BCE, Aristoxenus of Tarentum had indeed noticed this, according to Helmholtz, and had attempted to construct a scale by readopting the division of strings in harmonic ratios. This problem was taken up later and much developed by Gioseffo Zarlino, who was at the origin of what we might call the physicists' scale. Each note was obtained from natural harmonics of the fundamental transposed to the octave C1-C2, as with Pythagoras' scale. It goes without saying that for Helmholtz the natural scale was better than Pythagoras' scale, being based on all natural harmonics of the same tonic note rather than on the arbitrary base of just one, the fifth. This physicists' or natural scale, generated from all harmonics, differed from that of Pythagoras, generated only from the second harmonic of C0 (G1 from C0), in a number of technical points that we need not discuss here in detail. However, in relation to the tonic note C, E natural was 5/4 times higher, whereas for the Pythagorean E this distance was 5/5 x 81/80 (the "Pythagorean comma"). The latter was thus a little higher than E natural, and it was audible even to the untrained ear.

The Equal Tempered Scale: Werckmeister's Equation

But history did not stop at Zarlino. Indeed, a serious problem appeared during the sixteenth and seventeenth centuries with the development of harmony and melody. Increasing complexity of musical compositions made the musician change tonality, and thus the tonic note, more and more frequently in the same piece, usually with disastrous results for the correctness of the sound. If the piece was written in C major, for example, with C as the fundamental or tonic note, a change to D major after a few bars, with therefore D as tonic, imposed a slight modification of the pitch of most notes of that musical phrase in relation to the new tonic. The sixth A, for example, had a value of 5/3 in the natural scale of C major but of 3/2 in the natural scale of D major, in which it was considered a fifth. The A in D major was thus higher than in C major by a Pythagorean comma and necessarily sounded false on an instrument tuned to the natural scale of C major.

The result of changing the tonality of a string instrument implied each time either retuning the instrument, which was unrealistic, or changing the position of the fingers on the strings. For wind or keyboard instruments, one had to accept false notes unless someone could invent a new type of scale that remained "just" despite the changes of tonality. Andreas Werckmeister,[27] a German organist and theologian, proposed exactly that elegantly but totally arbitrarily: It was the equal tempered scale. In this ideal scale, the octave was divided into twelve equal semitones. On a piano keyboard, two white keys with a black key between them have an interval of a complete tone or two semitones. A white key and an adjacent black key have an interval of a semitone, just as for two white keys not separated by a black key. Each note could therefore be played, whatever tonality was chosen, without needing to modify its pitch, as in Pythagoras' or natural scales. For instance, A would not be made false if one changed from C to D major.

But how could one divide the octave into twelve equal parts? Werck-meister's solution was simple as long as one could reconcile the irreconcilable: Octaves followed suit in a ratio of two and fifths of 3/2. He observed that a succession of twelve fifths covered approximately the same extent of the sound spectrum as seven octaves, with the last B sharp fifth being distant from the last C by 74/73, however—that is, the fifth of a semitone. Not surprisingly, there was a slight difference between the results of progressions by octaves and by fifths because a power of two could never be the same as a power of three. To untie this Gordian knot, Werckmeister set the audacious equation: 12 fifths = 7 octaves. To solve the equation, he divided the ratio 74/73 by 12 and shortened each fifth by this twelfth, which amounted to spreading the effort over the ensemble of the fifths of the progression and thus bringing the B sharp to the same pitch as the last C of the progression by octaves. So he transposed each corrected fifth to the basic octave and obtained a scale in which all the semitones were practically equal.[28]

This equal tempered scale was adopted quickly by his contemporaries, notably Johann Sebastian Bach, who was able to cover all possible tonalities in his *Well Tempered Clavier*.

Helmholtz's Incredible Harmonium

However, one cannot say that Helmholtz shared this general enthusiasm for the equal tempered scale. He stated, rightly, that the fifth (G), but especially the third (E) and the sixth (A), of the tempered scale of C major were false in relation to the corresponding intervals in the natural scale. Further: "The dissonance due to harmonics appears a little gentler in equal tempered thirds than in Pythagorean thirds, but their combination tones are much more unpleasant ... they sound discordant and false."[29]

To support his statement and demonstrate the degree to which the natural scale was more just than the tempered, he conceived a project to construct a harmonium with two keyboards and reeds tuned so that if one changed register one could hear the same chord as either natural or tempered. The instrument was extremely complex because it had to modulate all tones in natural intervals. He concluded: "The musical effect of the natural scale differs noticeably from that of the tempered or the Greek system of fifths. The major chords of the natural scale ... offer a very full and saturated harmony; they flow in peaceful waves without trembles or beats. Compared with them tempered or Pythagorean chords seem harsh, troubled, trembling and irregular."[30]

A rather malicious tradition relates that his great friend, the violinist Joachim, had not been favorably impressed by the unaccustomed sonority of the harmonium, but he did not dare tell him. On the contrary, in a letter sent by the physiologist Carl Ludwig, Helmholtz learned that Ignaz Moscheles,[31] a rather famous pianist and composer of the period, was enthusiastic and that his preference was also for the natural scale.[32]

The Grandeur and Constraints of an Empiricist

Helmholtz warned us in his introduction to his *Physiological Basis for the Theory of Music* that he intended to bring physical and physiological acoustics together with musical theory and aesthetics. We must admit that he kept to his word. During his research marathon, he neglected nothing that could unite these specific and complementary disciplines.

We might well be surprised at the diverse facets of his personality. He was an unsurpassed technician with the ability to conceive and construct precision laboratory apparatus perfectly adapted for his research, a theoretical physicist and excellent mathematician often forecasting the results he was to obtain experimentally, and a competent anatomist and physiologist with intuitions sometimes preceding by decades the moment when they could be tested experimentally, as well as an expert and enthusiastic connoisseur of the history of music and musicology. In addition, he had a marked sense of the theatrical because he manifestly felt a certain pleasure in surprising his contemporaries by presenting himself, the specialist in vision, after only a few months of extra work as he who had discovered the "true and sufficient causes of consonant and dissonant behavior of musical sounds."[33] With his will to unify so many different scientific disciplines in a coherent entity, he proved once again his veritable gluttony for science and knowledge. But a strange thing occurred: In this haughty vision encompassing acoustics, the ear and music, there was once again a major absentee: the brain. The reasons are doubtless the same as those we evoked in chapter 9 related to his work in physiological optics: a reluctance to equate anatomical and psychological processes in absolute terms.

An essential point in Helmholtz's methodology was the importance he attached to the concept of the model as a preliminary theoretical necessity before any experimental study. We have the impression that, at the time he left Königsberg for Bonn, although he had only just ordered his research apparatus and his laboratory was not yet ready to receive him, he had already made all the calculations for the experimental scenario that he had conceived and which was to lead him to a mechanistic explanation of auditory sensation. He also used numerous analogical models, reproducing with instruments data from the physicomathematical world: various sirens, tuning fork sound generators, resonators, and a harmonium. But above all, he conceived vast and ambitious theoretical models from two extremes, on the one hand postulating the coherence of the physical world in mechanical terms, and, on the other hand, perception as a result of unconscious interference at the sensory level. The ear was the obvious strategic interface: a musical ear.

But let us examine that more closely. All his life, Helmholtz proclaimed himself a confirmed empiricist. He was indeed, but with some important nuances imposed by his fidelity to Kantian principles, such as a priori time and space, which he "enriched" with physiological concepts such as the legitimization of space through movement. Furthermore, he sought in his empiricist approach to knowledge an objective truth, a concept of Nature defined in terms of elementary forces, and he demonstrated that he was the only possible one who was compatible with the facts.[34]

In his work on music, he systematically applied his own doctrine. He described the physical basis of sound and harmonics, timbre, beats and combination tones, and consonance and dissonance. He analyzed and described quantitatively the tonal complexity of human vowels and musical instruments. Because complex sound could only be analyzed by using resonators, he found it obvious that the ear, which possessed structures—pillars and hair cells—that were susceptible to play such a role, was the detector and judge of sound quality. The ear recognized timbre and whether a chord was consonant or dissonant, and thus pleasant or unpleasant. The history of music as later described by Helmholtz revealed the irresistible rise of consonant harmony over 2,500 years as a musical value *par excellence*. In the same way, he explained why the strict development of the rules of melody was necessary for the beauty of consonance in a succession of different chords. As to scales, they had to be natural because these were less false than any others and relatively free of possible dissonance between harmonics and combination tones.

As we can see, the ear played a capital role because it analyzed sound rigorously and separated consonance and dissonance implacably. At certain moments of his research, Helmholtz was probably tempted to take the step of believing that the ear "knew" music because it detected pleasant sounds. He did not take that step, however. On the contrary, he insisted that the ear only intervened for the benefit of the senses, so furnishing an important means of reaching the goal of the pursuit of beauty. We must admit that, in the domain of aesthetics, Helmholtz often gave the impression that, consciously or not, he was indulging in a ballet in which he pirouetted sometimes metaphysically, sometimes mechanistically, but with a preponderant role for the ear.

The psychophysicist Gustav Fechner [35] called aesthetic all that entered through our senses to give immediate pleasure or displeasure. Thus, aesthetics for him was simply related to the concept of pleasure—the science of the pleasant or the unpleasant. Helmholtz clearly tended toward mechanistic aesthetics stating without exaggerated modesty that in many recent compositions dissonant chords were the rule and consonant chords the exception. No one should doubt that the opposite should be the case and that the prolonged use of daring and offensive modulations risked the disappearance of the sense of tonality. They were, he said, annoying symptoms for the future development of the art. Considerations of mechanisms of instruments and facilities of execution tended to take precedence over the natural exigencies of the ear.[36]

In his opinion, great music had probably reached its apogee with Wolfgang Amadeus Mozart, Ludwig van Beethoven, and perhaps Felix Mendelssohn, and it is significant that at the time he was writing his treatise, Frederic Chopin and Robert Schumann had been dead for several years (1849 and 1856, respectively) without having been, it seems, explicitly appreciated by our grand physiologist. When the fifth edition of the *Theory of Music* was published in 1895, he had had the opportunity to listen to Johannes Brahms and Franz Liszt and perhaps even hear the first works of Gustav Mahler, Anton Bruckner, and Claude Debussy, but he never spoke of them. Although an assiduous habitué of Bayreuth and the Wagner family, the history of quality music seemed to have ended for him at the beginning of the nineteenth century.

But what did he really think when listening to the *Magic Fire Music* during Brünnhilde's enchanted sleep, with its succession of chords of which the tonalities were so far removed from each other or Tristan's theme with its incredibly audacious harmonics for the time? How would he have reacted if he had heard Arnold Schönberg's piano suite in 1921, in which he used a dodecaphonic scale where there was no longer a tonic or fundamental note and where each of the twelve notes of the octave were considered as equally important reference points? The ultimate experience for him would doubtless have been the astonishing performance of the contemporary composer Giancinto Scelsi, who, in his four pieces for orchestra, only used a single note during several minutes, always the same but with changing rhythm, played by all the instruments, sometimes pure, sometimes animated by intentional beats provoked by slight differences in pitch between one instrument and another. This is real music that demonstrates well that musical genius cannot be reduced to a few physical or physiological mechanisms.

Helmholtz only partially gave in to the almost Faustian temptation to create a set of essentially scientifically based aesthetics because he was all too conscious of the complexity of musical perception. Perception, which enabled one to recognize a vowel, a musical instrument, or the composer of a given piece of music, was explained for him in psychological terms, and he never accepted to reduce it to purely physiological mechanisms. However, one must credit him with having attempted to obtain for physics, and the ear, a role in the euphonic perception of music that no one had ever dared imagine before him. He left us with the results of one of the most brilliant scientific enterprises of the nineteenth century, rendered even more attractive by its author's view of a scientific utopia.

As to the beauty of a musical composition, and the greatness of the creative genius in which he believed most sincerely, he explained it in metaphysical terms near the end of his treatise: "In us emerges a sentiment of rationality of the masterpiece which extends far beyond what we can encompass at that moment, and of which we can see neither boundaries nor limits. Mindful of the poet's words 'You are the equal of the spirit that you conceive'[37] we feel the spiritual forces that have been at work in the artist and that are far superior to our conscious thoughts."[38]

So Helmholtz's dilemma appeared to be resolved, but his struggle between Faustian temptation and mechanistic aesthetics, and his devotion to a metaphysical definition of beauty, persisted.

Conclusion: The Wisdom of Alexander von Humboldt

My essay on the cosmos is the contemplation of the universe founded on rational empiricism, that is to say on the ensemble of facts recoded by science and subject to the operations of comprehension which compare and combine.
—Alexander von Humboldt[1]

Nostalgia

During his life, Helmholtz was hard on himself and exigent on others. He never hesitated to face up to polemic with those who did not share his own vision. Yet in the twilight of his life, basking in honors and social success, he could even be nostalgic. One of his last photographs, taken in 1894 shortly before his death, shows him standing giving a lecture, surrounded by figures and formulae on the blackboard, but with a light in his eyes that expressed a certain moral fatigue, if not sadness.

It is true that the superb new research institute that he had just created with the help of his friend Siemens, the famous industrialist and his daughter's father-in-law, his eminent role of Privy Councilor to Kaiser Wilhelm, and his friendship with a host of great minds such as du Bois-Reymond, his most faithful friend, could not make him forget that times had changed. A lot of water had flowed through the Baltic, the Rhine, and the Neckar since his celebrated research on their respective banks. Life science had not ceased to progress the day he had abandoned it to study natural sciences, such as physics, electricity, and thermodynamics. Further, his detractors had not laid down their arms. There was Hering, the inneist and intransigent successor of Purkinje, and his student Wilhelm Wundt, who had left him on bad terms to pursue a brilliant career in experimental psychology. Then there was Ernst Mach, who was at first his thurifer in publishing his *Introduction to Helmholtz's Theory of Music*[2] and who later became one of his most virulent contradictors.

Even William James, the famous American psychologist who although admiring him enormously, launched a number of bitter attacks that had hurt him.[3] It is possible that the rise of these criticisms contributed to making him leave physiology for physics, convincing him that he had done the maximum to make his views understood. Indeed, it was characteristic of him that he reedited his books several times throughout his life without attempting to adapt them to new fashions except by adding philosophical comments. His works were set permanently in stone.

During this time, the study of the function of the nervous system had evolved considerably. Charles Sherrington was at the beginning of his career, and his interpretation of behavior based on a hierarchy of reflexes left little room for psychology as favored by Helmholtz. In a letter to Franciscus Cornelius Donders in 1868 about eye movements, Helmholtz stated: "Reflex movements as we can now define them are everything in physiology that we cannot explain. They are the residue of exaggerated material metaphysics from which people must be recalled, back to the facts."[4]

Ivan Petrovitch Pavlov,[5] the star of Russian physiology, despite the considerable number of Helmholtz's students in his country, attempted shamelessly to adopt the latter's views by incorporating in his own theory of conditioned reflexes the mechanisms of unconscious interference and apprenticeship inherent to his theory of perception.[6] We should recall that Helmholtz was not viewed in a good light by Lenin, who accused him of Kantism, subjectivism, and agnosticism.[7] On another plane, Freud had seized on the unconscious and had endowed it with an imperial role in directing one's thoughts and actions to the exclusion of self. Finally, many philosophers consecrated their rupture with science through an exclusive and often intolerant idealism, a real drama for the progress of knowledge and learning.

Helmholtz regretted that he had not always been liberal enough with his own students, and the departure of Wundt, who was tired by the overbearing authority of his mentor, had left him with a bitter taste. Helmholtz also regretted a serious lack of judgment concerning Charles Darwin, although the latter's research went in the same direction as his own, demonstrating the crucial importance of adaptation and apprenticeship in the evolution of species. Instead of appreciating the work of the great Darwin, he had been suspicious of this "left winger." In his speech on the occasion of his installation as Rector of Berlin University in 1877, devoted to academic freedom in German universities, he did not hesitate to suggest as proof of this liberty the fact that one could "expound the most extreme consequences of materialist metaphysics, and the most audacious speculations on the basis of Darwin's theory of evolution, just as freely as the most extreme deifications of pontifical infallibility."[8] He subsequently made amends by eulogizing him on many occasions,[9] but he had nevertheless missed the occasion of establishing an obvious complementarity between Darwin's ideas and his own.

The dazzling development of natural science at the end of the nineteenth century was a foretaste of the discovery at the beginning of the twentieth century of relativity, the atom and the quantum, and had destroyed the illusion that knowledge was the same as ultimate truth. Helmholtz was well aware of this mutation in ideas because, since he had left Heidelberg for Berlin, several of his certainties had insidiously mutated into uncertainties.[10]

The Manes of Goethe

It was in this state of mind that he received an invitation, as unexpected as that of forty years previously in Königsberg, to give a lecture on Goethe to the general assembly of the Goethe Society in Weimar. He accepted with pleasure and titled his talk "Goethe's Presentiments about Future Ideas in Natural Science."[11] This talk was important because, without abandoning any of his convictions and without retreating on the essential criticisms that he had expressed earlier about the poet's scientific errors, he explained, on the contrary, what united Goethe and him, their true complementarity in the mutual quest to understand nature. It was a sort of moral and even spiritual testament that revealed perhaps a certain disenchantment related to age, but in which his enthusiastic conviction that art and science worked in the same direction dominated incontestably.

Creative Intuition

Always faithful to Kant, Helmholtz began by recalling that only the sensory organs opened the doorway to consciousness, and that it was thanks to the concept of intuition elaborated by the philosopher that scientists and artists had found solid common rules for acquiring their respective knowledge. Kantian intuition meant seizing in a single act the raw results of sensory representation in the context of an a priori form of space and time. Fortified by this certainty of obtaining knowledge safely, the scientist pursued observation and inductive research to investigate the laws of nature. However, these laws, expressed in words, were mere hypotheses needing verification by confronting them with facts. It was only by proceeding with care that one could assume them to be correct and legitimate at least under given experimental conditions. Then the scientist was like a prophet or magician because he acquired power over nature and became able to predict certain phenomena.

He also said that, using intuition, the artist could equally achieve as true a vision as the scientist of the complex phenomena of nature and the human mind. For example, tragedy made us aware that, in his place, we would have pursued the same terrible actions and passionate behavior of the hero. This metaphor revealed the true nature of the mind, its logical structure, and its closeness to the emotional reactions of a hero. Such a concept of the mind was certainly not the result of philosophical thought, which would doubtless have been a handicap, but that of artistic intuition.

The intuition of the scientist and that of the artist had the common feature of appearing instantaneously independent of any thought. We might consider how we easily localize a familiar object in space intuitively, thanks to memory. It was the same for artistic intuition, which was neither the result of thought nor expressed in words, but which appeared instantaneously, full of life and affectivity. Art, like science, could convey and spread truth. A painting, however, must never be like a photograph of a given subject but rather the representation of an ideal, and the artist must be ready to refuse to reproduce nature faithfully if necessary for the expression of beauty. What was pleasing was not necessarily beautiful, but musical consonance or balanced luminosity in a landscape provided a favorable environment for perception by our soul.

Helmholtz continued that memory was essential for intuition, especially for the artist who was trying to represent on a single canvas the waves breaking in the sea or lightning streaking the sky of a landscape plunged into the darkest night. In science, memory also played an important role, and one could cite as an example Goethe's intuition of his vertebral theory of the skull simply by seeing the scattered debris of a sheep's skull on a Venetian beach.

Goethe's Scientific Intuition

According to Helmholtz, Goethe's scientific glory reached its peak in morphology. His artistic way of observing phenomena led him to favor common structural patterns for all animals and plants. However, this conviction, which he persuaded a number of his contemporaries to share, did not stand up to Darwin's discovery of the major morphological changes that animals could undergo spontaneously even in a few generations. He also failed in his study of color; lacking adequate optical apparatus, he was unable to describe the true fundamental colors.

However, he added, Goethe had been quite right to seek to reduce the multiplicity of his observations to a single primeval phenomenon. For example, he was opposed to the abstraction of nonintuitive concepts, as was rife in theoretical physics: force without matter or matter without force. One had to recognize that these ideas, legitimate in the seventeenth and eighteenth centuries, had been the source of considerable confusion, such as animal magnetism or vital force, which were simply forces without matter. Unfortunately, Goethe had become lost in polemic against physics, and it was all the more regrettable that modern physics had in the main adopted the essentials of his ideas. It was just as regrettable that the poet was not aware of Huygens' wave theory, which, in his hands, would certainly have become a primeval phenomenon. Indeed, mathematical physics drew its force not from Goethe but from Huygens. In this respect, Goethe's true spiritual heir was Michael Faraday, an unprejudiced scientist who detested abstract concepts that he did not know how to manipulate.

The law of gravity was another example of the primeval phenomenon only accessible to observation. Newton had formulated his famous law retaining only what could be measured: mass, speed, distance, and acceleration. He had the genius not to add what was beyond observation except in terms of similitude, notably by saying that the planets moved *as if* they were attracted to each other by some force. It is remarkable to realize that Goethe did not hesitate to use the term of similitude in an eulogious study that he devoted to Francis Bacon in his *Theory of Colours*. He also said, "I have accepted that I shall never understand the primeval phenomenon but I remain anxious for I do not know if its cause is the limits of the human mind or that of my own ignorance."[12]

The Common Objective of the Poet and the Scientist
Goethe did not really like Kant's *Critic of Pure Reason*, preferring the *Critic of Judgment*[13] because of the boundary, artificial according to him, that Kant had established between data from observation and a priori space and time. But Helmholtz claimed to have demonstrated that in a theory of space, these a priori principles were in reality physiological and therefore accessible to experimentation, which would doubtless have pleased the poet.

Helmholtz continued: There was no more similarity between a sensation and an object than between a word and the object it named because sensation was merely a sign of the effect of the object on our senses. But there was an exception for whatever occurred over a period of time, he added, because in this case and this case only, the sensation was already a symbol of *reality*. Goethe had also understood this because in *Faust* he put into the mouths of the mystical choir describing the holy souls facing the eternal Truth: "All transient things are but a parable: the inaccessible here becomes actuality; the ineffable here is achieved; the Eternal-Feminine draws us onward."

Alles Vergängliche
Ist nur ein Gleichnis;
Das Unzulängliche,
Hier wird's Ereignis;
Das Unbeschreibliche,
Hier ist's getan;
Das Ewigweibliche
Zieht uns hinan.[14]

This could be interpreted that anything occurring over a period of time and that was perceived by the sense organs could only be taken as a symbol, an image, or a metaphor.

Helmholtz did not stop his comparison there. He said understanding a natural law was inductive in origin and so always remained incomplete. The poet felt the same as the scientist; by making the mystic choir proclaim that the inaccessible became fact, he transformed our sentiment of helplessness into anguish. Goethe added that we could only know the ineffable, which could not be expressed in words, in the form of an artistic representation: "The ineffable here is achieved; the Eternal-Feminine draws us onwards." For sanctified souls it was a reality, but for scientists the footsteps left by the poet became more and more subtle and difficult to follow. One must recognize that even if the objective remained the same, the connections between the views of the scientist and the intuition of the poet were often blurred and difficult to justify.

Action to the Rescue of Knowledge
In his despair over not achieving knowledge of the world and so ensuring his power over reality, Faust first sought a solution from the beginning of the gospel of St. John: "In the beginning was the word." Because the word was nothing more than a sign that allowed its meaning to be known, the word thus became a concept or, if it referred to a current event, a law of nature. If this law was effective and resisted over time, it was called force. This passage from word to concept to force could not satisfy Faust, who then cried, "In the beginning was the deed." For Goethe it was less

a question of understanding the origin of the world than finding the pathway to the knowledge of truth.

As a counterpoint to this scene from *Faust*, Helmholtz could not resist recalling the vain efforts of philosophers to establish the existence of reality as long as they restricted themselves to passive observation of the outside world. They could not go beyond the world of images and did not understand that human activity, under voluntary control, was an inexhaustible source of knowledge. Our sensory perception, he continued, was merely a sign language concerning the outside world, and we could only decode it if our actions met with success, this enabling us to distinguish impressions that could be modified by our actions from those that were independent of our will. We gained knowledge of reality through action, and the greatness of sensory physiology was to demonstrate how we could do this successfully.

He concluded his lecture on Goethe by recalling that he had reached the highest summits of his art when the problems he tackled were solvable by poetic divination using intuitive images. On the contrary, he had failed when only the conscious use of the inductive method could have helped him. Fortunately, in important questions concerning links between reason and reality, he always maintained his solid good sense, which protected him from aberrations and enabled him to obtain in all security visions that reached the limits of human comprehension.

The reader might wonder whether it was really useful to concentrate at such length on a lecture that was perhaps merely a mundane ceremony in which an old man consecrated another, who had been dead for several decades, after having been for a long time his posthumous, but scathing, adversary. That would be an interpretation that might be justified on the part of a disillusioned mind. On the contrary, we might see one of those moments of grace in the history of mankind and his thoughts when lucidity of mind took priority over the susceptibility of sentiment. These were two very different men—different in every respect, even living in different eras, but incontestably animated by the same passion to know and to understand, despite all the obstacles, what is nature, what is life.

The fact that their lives only just overlapped in no way detracts from their controversies. It is true that divergence of view between contemporary scientists is often technical in nature and is usually resolved by later advances in science, whereas in the case of Helmholtz and Goethe, the problem was of greater magnitude and extended beyond their epoch, so much so that it concerns us today. The intemporal nature of the problem prompts us to ask the question, without particular incongruity: What would Alexander von Humboldt have thought of all this?

Alexander von Humboldt between Enlightenment and Romanticism

We, of course, do not know what Alexander von Humboldt would have thought because he died in 1859. However, we could wonder how the sage of Potsdam would have reacted to the lecture in Weimar by his protégé Helmholtz.

Humboldt was an heir of the Enlightenment in terms of his knowledge, his encyclopedic culture, and his permanent concern for tolerance, justice, and equality among men. He was welcomed in triumph in the United States after his South American expedition, but he did not hesitate to reproach President Jefferson for the maintenance of slavery and the slave trade. What is more, he appreciated the fine arts and stimulated the creativity of many artists thanks to sketches brought back from his travels. Few personalities would have been better placed to appreciate the respective merits of such contrasting men as Goethe and Helmholtz, both of whom he had known well.

Humboldt struck up a friendship with Goethe, and both men had a deep respect for nature. He had been a passionate explorer of the unknown lands of Central America, describing in detail geography, waterways and climate, mountains and volcanoes, fauna, flora, and population. He also discovered the richness of the interactions between these diverse factors: man as tributary of his environment, but also sculptor of his own landscape, privileged actor in the equilibrium between animal species. For this ecologist ahead of his time, nature was a vast network where all activities enjoyed reciprocal relationships.[15] He was convinced, as appears from his *Cosmos*,[16] that his work had nothing to do with speculative philosophy. He wished to look on the universe as an empiricist, submitting his scientific observation to intellectual scrutiny. As an anthropologist, he also sought to inscribe his discoveries about the physics of the Earth in the pages of the story of the cosmos in human culture because he was convinced that the history of views of the world did not reveal repeated oscillations between truth and error, but rather the principal moments of a progressive road toward truth, toward a true representation of terrestrial forces and the planetary system.

Humboldt's certainty of an occult influence in the evolution of the world had to please Goethe. Indeed, at the time of a memorable meeting in Jena, he found in the brothers Alexander and Wilhelm von Humboldt sympathetic ears for his ideas on comparative morphology, particularly his studies in osteology, the germ of a global comprehension of life. In fact, he often met Alexander with whom he had a warm relationship. On the occasion of one of his visits, he did not hide his pleasure, confiding in Eckermann: "On whatever point you approach him, he is at home, and lavishes upon us his intellectual treasures. He is like a fountain with many pipes, under which you need only hold a vessel, and from which refreshing and inexhaustible streams are ever flowing."[17]

For his part, Humboldt held Goethe in great esteem because he admired his global vision of the world in flux, his exceptional gift of observation, and his encyclopedic knowledge. He was doubtless less convinced by the scientific value of his research and certainly regretted his irrational rejection of Newton's theories. We must remember that Humboldt was a vulcanist and Goethe a neptunist (see chapter 7). Indeed, the latter expressed his abject contempt in a letter to Humboldt in support of a friend of his, a Polish pianist: "Because you belong to those scientists who consider that all things were produced by volcanoes, I am sending you a female volcano who submerges and consumes everything."[18] However, he heartily embraced his vision and aesthetic intuition because he believed, like him, that aesthetics were the motor for a global view of the world.

Helmholtz might not have had the opportunity to know him had he not also been so close to Johannes Müller. Few people had succeeded as the latter did in penetrating the labyrinth of natural philosophy and romantic thought and still manage to achieve an integrated concept of science, away from the clutches of any metaphysical consideration. Humboldt had been sympathetic toward natural philosophy at the beginning of his career, but he had soon fustigated its excesses, denouncing the Saturnalia of a purely abstract philosophy. For Müller and he, knowledge depended on the preexisting, a reintegration into the bosom of a superior order that, however, Schelling was far from wrong in stating that all surrounded us in the fullness of its significance.[19] Humboldt was an experimentalist and had moreover intervened personally at the beginning of his career in the controversy that opposed Allessandro Volta and Luigi Galvani, stating that in the frog neuromuscular preparation, a contraction could happen under the influence of a nervous "galvanic fluid" originating in the living tissue and resulting from modifications in tissue chemistry.[20] On his advice, this research was later followed up in Müller's laboratory, in fact by Emil du Bois-Reymond.

Humboldt had known Helmholtz since the beginnings of his research and immediately supported him for his audacity in proclaiming the unity of physical and life sciences under the influence of mechanical physics.

Later he was full of admiration when his protégé succeeded in measuring the speed of conduction of a nerve impulse and so demonstrated, contrary to conventional wisdom, that this speed was far from being that of light and was in fact only a few tens of meters per second. At that moment, he became conscious of the fact that this discovery had sounded the death knell of psychophysiological parallelism because both time and nervous conduction were thus implicated in psychic processes such as sensory perception and motor commands, which had been felt to be instantaneous. He presented these results at the Academy of Sciences in Paris.[21]

We may then think that Humboldt would certainly have been very happy at this sort of posthumous reconciliation between Helmholtz and Goethe. Helmholtz provided numerous proofs that science must be stringent and that this was the price of any progress toward an understanding of the mechanisms of nature, even if he had occasionally surrendered to the sirens of metaphysics. That systematic observation was an indispensable arm for science was obvious, and Goethe used it marvelously well. However, his error was to have refused to analyze what he observed and to have rejected the use of mathematics and experimental apparatus that might have disturbed his global vision of nature. Goethe had therefore never been a scientist in the strict sense of the word. That Helmholtz recalled it in his lecture would certainly have pleased Humboldt.

But what he would have found new and important was the fact that Helmholtz further recognized that the poet contributed quite specifically to the acquisition of knowledge. In his opinion, Goethe's art not only enabled him to describe man's states of mind with a veracity that was inaccessible to the philosopher or empirical scientist, but it had been possible for the father of Faust to disseminate universal principles of knowledge, such as Helmholtz had discovered in his physiological studies thanks to the power of his artistic intuition and his metaphorical visions. That sensory awareness of the environment depended on voluntary activity had therefore become a reality for both poet and physiologist. Thus, each at his own level of competence accessed the same package of knowledge in complementary fashion.

In his reconciliation with Goethe, Helmholtz showed proof of much humanity in admitting the limits of his own conquest of scientific knowledge, and recognizing in the artist, and in his aesthetic approach to man and nature, a form of knowledge that was unique to him. A thirst for knowledge honored humankind, and science was indispensable for quenching it. However, when this knowledge respected that of the artist, it became wisdom.

Postface

The lives of great men of the past often seem like a *trompe-l'oeil* to those who wish to scrutinize them. Their characters seem fixed for all time; their prestigious works, analyzed once for all, remain in our memory like incunabula that we respect but do not open anymore, but they take on another aspect as soon as we get closer to them and we see them in a different perspective.

This is what I experienced during the writing of this book. At first my characters were like statues, but then they came to life and gradually changed, sometimes in unexpected ways. We have reached the climax of this metamorphosis and are about to end, and there is no reason to want to summarize what has been said and risk fixing once again the images that have emerged from the past in their new light. However, it seems fitting to highlight certain aspects of Helmholtz's personality and career that may have been rather neglected in the main text but that could be useful in our analysis of him more than 100 years after his death.

Helmholtz adhered sincerely to the three Greek pillars of experience of the "honest man": truth, goodness, and beauty. The search for truth was at the center of his scientific preoccupations. He did not have to bother himself too much with goodness because he was a man of duty and, as all true Prussians should be, highly disciplined. What caused him a problem was beauty because he did not know too well how to situate it in relation to science. His beauty was that of Plato, the same for everyone and everywhere. In his eyes, it probably deserved the same status as Kant's a priori space and time. Of course he showed that science was at least capable of helping the musician in the shape of his discovery of the essential role of consonance and dissonance in musical pleasure. As for painting, the pointillists Seurat and Signac would have been able to bear witness to the utility for their art of his research on color. But he expected

more of science: if not a definition of beauty, at least an explanation of
the physiological laws of perception.

Faithful to his philosophy of reducing if possible all observed phenomena to a single explanation, he gradually acquired up to the end of his
career the conviction that the artist's perception was built up in the same
way as the scientist's—by unconscious sensory inferences. The intellectual processes were the same in both cases depending on memory and
imagination by association. In the realm of the unconscious, the processes of reasoning took place as in the conscious mind, and, for artists
as well as for scientists, induction consisted of detecting in the flow of
sensory information those elements representing a coherent whole that
made sense for an individual. The difference between the artist and the
scientist was that truth as seen by the artist was not the copy of a single
object but the representation of an ideal.[1] When he painted, an artist did
not copy nature but transformed it in order to show what this aspect of
nature contained in terms of universal and ideal. Helmholtz therefore
proclaimed himself in favor of a classic ethic in which beauty emerged
from the perception of the ideal despite the imperfections of the particular,[2] and this indeed corresponded perfectly to his own artistic taste. For
Helmholtz, "the ultimate secret of artistic beauty, that is to say the wonderful pleasure that we feel in its presence, depends essentially on the
sentiment of lightness, harmony and vitality of our imagination ... of
which the same goal is to make us more aware of so far hidden laws, and
to enable us to contemplate the ultimate depths of sensitivity of our own
mind."[3]

Helmholtz's reconciliation with Goethe's ideas that we discussed in
the *Conclusion* was far from being pretence. It stemmed from his certainty that the physiological processes activated in perception were identical in artists and scientists. Indeed it had become evident to him that
beauty emerged from the perception of truth and that up to a certain
point the paths of artist and scientist were of necessity common ones.

Goethe's convictions about the relationships between beauty and
truth were, however, noticeably different. Goethe never accepted being
a slave of one or other school of philosophy, and this explains why his
concepts of aesthetics may seem complex and even contradictory. It was
as if there were two personalities in him: one classic and the other romantic. The classic Goethe stood in wonder, as Helmholtz would have, before
the Laocoon sculpture, of which he admired the gestural and anatomical
perfection that bore witness to the vast knowledge of the artist and yet

was far from being an imitation of nature.[4] For him, Greek art was the only one to which we owed an eternal debt; for all the others, we had to lend something. The other Goethe, a romantic and natural philosopher, had been strongly influenced by the English philosopher Anthony Ashley Cooper, Earl of Shaftesbury,[5] for whom aesthetics was the keystone of philosophy: All beauty was truth, and the truth of the cosmos had its place in the phenomenon of beauty.[6] Goethe endorsed this when he defined beauty as a manifestation of the secret laws of nature, which would have remained eternally hidden from us without it.[7]

This brief perspective on the aesthetic concepts of Helmholtz and Goethe provides us with two messages. Like Janus, Goethe did not conceal his taste for classicism while cultivating his romantic vision of nature, the beauty of which increased his thirst for a global understanding of the world, irrevocably linking his poetic and scientific undertakings. As to Helmholtz, who was an eclectic amateur of art and a highly talented musician, but with an essentially classical culture, he had the enormous merit of attempting a physiological approach to the phenomenon of aesthetics, prudently but with genius in his research on music, later audaciously in his work on painting and in his last lecture in Goethe's honor.

According to the Prussian minister Heinrich von Mühler in 1870, Helmholtz was as prestigious a scientist as Alexander von Humboldt. This notion shows us that, in the second half of the nineteenth century, Helmholtz was considered the principal spokesman for science in Germany, just as Humboldt had been before him and as his student Max Planck would be at the beginning of the next century. Helmholtz derived this exceptional authority above all from his scientific competence but also from his cultural and artistic influence. In an important work, Bouveresse[8] devoted much space to Helmholtz's ideas on perception, which illustrates well how as a scientist he also occupies an important place in the reflections of modern philosophers. However, it is of interest to note that the famous treatise of Merleau-Ponty in 1945 never cited the great physiologist's work.[9]

During the last twenty years of his life, Helmholtz devoted much time to making science known widely. In this undertaking, he not only included physical science, essential as it was in his eyes, but also human science, which he considered an indispensable complement to the former. Moreover, he was convinced, as we just discussed, that art and natural science enjoyed a close reciprocal relationship. Like most German academics,

Helmholtz was convinced that a systematic, global study of all these aspects of science, despite their different methodologies, was the only means of achieving a high level of culture and civilization.[10]

In his lectures to cultivated audiences, he constantly emphasized what for him was at the heart of natural science: (a) its essentially empirical and experimental character, (b) the fact that observation alone did not achieve scientific status unless it enabled one to deduce laws and establish the causes of phenomena, and (c) the reciprocal interactions between observed facts and theory, because the natural philosophers had committed the unpardonable error of wanting to make deductions before establishing their theories by induction. Nevertheless, Helmholtz did what he reproached the natural philosophers for when he admitted without further commentary the specific nervous energy of his mentor Müller (chapter 9). He did the same when he relied on Kant's metaphysics in his work on the conservation of force and perception (chapters 5 and 9). Finally, he highlighted the independence of the laws of causality that operated in nature beyond our perception, our thought, and our will.

By attaining a high level of scientific culture, he estimated, his fellow citizens would have a clearer view of their own place in the natural world. Thanks to physiology and anatomy, they would better understand the functions and needs of their body and, when sick, would seek aid from scientific medicine rather than from charlatans and superstitions. A better understanding of the origins of man, made possible by Darwin, would qualify then as citizens of a planet with a natural history and an ever-changing future. As to discoveries due to research in physics and chemistry, they would encourage people to use them in industry, where technology would develop for the greater good of humankind. In harmony with human science, natural science would thus ensure the supremacy of man and his intelligence.

An exemplary scientist and author of important major research projects, Helmholtz took great care to disseminate his results and draw conclusions from them not only for his own glory, which was human, and for the benefit of other scientists for whom he was the spokesman, but also to spread to his compatriots his ideas on scientific progress and its promises for the future. Cahan asked, quite rightly, if the three keywords *reason, empiricism,* and *utility,* which for Helmholtz were practically the watchwords of his life, did not make of him, at the end of his century, the last representative of the Enlightenment.[11] The idea is attractive. However, like Cahan, we might be surprised that with his unreserved faith in scientific progress, promising happiness and prosperity in its wake, Helmholtz apparently never perceived the dangers, already present in his day of poorly controlled technology, nor did he foresee the misdeeds that could be perpetrated in the name of science. Did he not realize that, if the acquisition of new knowledge was a duty for all, the exploitation of its consequences necessarily engaged the moral conscience of all?

Nevertheless, it remains true that, at the beginning of the third millennium, Helmholtz remains in our eyes a scientist of exceptional stature, whose scientific rigor and strict empiricism seem ever more exemplary. Neuroscience and cognitive science, as we call them today, owe numerous research domains to him, as well as attitudes. No phenomenon of nature, life, or environment left his encyclopedic mind indifferent. He believed he could reconcile science and philosophy, notably by thinking that Kant's a priori had in the last resort a physiological basis that would one day doubtless be discovered.[12]

An enlightened and enthusiastic melomaniac, he brought together science and art, one of the first physiologists to do so. Ambitious in his search for knowledge, he saw the united practice of physical science, human science, and fine art as the only way to ensure the harmony of the human mind in its quest for knowledge and wisdom.

Notes

Preface

1. 1857–1952.
2. Sherrington 1906.
3. Gregory 1970.
4. 1837–1921; Koenigsberger 1902–1903.
5. 1724–1804.
6. 1762–1814.
7. 1801–1858.
8. Helmholtz 1856–1866.
9. Helmholtz 1863.

Prelude

1. 1596–1650.
2. Descartes 1637.
3. Desné 1972, p. 83.
4. Goethe 1810.
5. 1792–1858, Koenigsberger 1902 I p. 6.
6. 1712–1786.
7. Im Hof 1993, p. 11.
8. Streidt & Frahm 1996.
9. 1699–1753.
10. Voltaire 1784; Streidt & Frahm 1996.
11. Koenigsberger 1902 I p. 7.
12. Desné 1972, pp. 83–84; D'Alembert 1751.
13. Adamov-Autrusseau 1972, p. 121.
14. 1623–1704.
15. 1711–1776.
16. Kant 1784.
17. Streidt & Frahm 1996.
18. 1775–1854.
19. 1770–1831.
20. 1778–1860.
21. 1749–1832.
22. 1744–1803.
23. 1733–1813.
24. 1759–1805.
25. Ludwig 1920 I, pp. 52–53.
26. Goethe 1810.
27. Goethe 1790, cited by Lacoste, 1997, p.51.

28. Goethe 1810.
29. Goethe 1810, pp. 130–131.
30. Lacoste 1997, pp. 221–227.
31. Cassirer 1923–1929.
32. Goethe 1790.
33. Lacoste 1997, p. 84.
34. 1642–1727.
35. Goethe 1810, p. 131.
36. Lacoste 1997, p. 107.
37. Lacoste 1997.

Chapter 1

1. Hörz & Wollgast 1986, p. 40.
2. Helmholtz 1877a.
3. Kaschuba 1995.
4. Helmholtz 1865.
5. Helmholtz 1863.
6. Koenigsberger 1902, I p. 289.
7. Heine 1836.
8. Cahan 1993b, p16.
9. Koenigsberger 1902, I pp. 29–33.
10. 1801–1858.
11. 1818–1896.
12. 1819–1892.
13. 1856–1939.

Chapter 2

1. Aromatico 1997.
2. 1736–1788.
3. 1708–1777.
4. Helmholtz 1877a.
5. 1548–1600.
6. 1632–1677.
7. 1646–1716.
8. Seidengart 1989.
9. Bruno 1591; Gandillac 1973.
10. Spinoza 1677.
11. Bouquiaux 1995.
12. Leibniz 1714.
13. 1801–1887.
14. Besnier 1993, p. 139.
15. 1744–1803.
16. 1752–1831.
17. Kant 1784.
18. Ludwig 1920, p. 67.
19. 1730–1788.
20. Adamov-Autrusseau 1972, p. 129.
21. Adamov-Autrusseau 1972, p. 131.
22. Herder 1773.
23. Fink 1991, p. 9.
24. Goethe 1773.
25. Heine 1836.
26. Fink 1991, pp. 12–14.
27. 1779–1861.
28. 1772–1801.

29. Alexandre 1963, p. xiv.
30. R Huch, cited by Alexandre 1963, p. xxviii.
31. Marquet 1993, p. 77.
32. Schelling 1797.
33. Tilliette 1973, pp. 972–973.
34. Cassirer, cited by Besnier 1993, p. 257.
35. Gusdorf 1985, p. 115.
36. Gusdorf 1985, p. 53.
37. Schelling 1799; cited by Tilliette 1973, p. 973.
38. Gusdorf 1985, pp. 43, 45.
39. Faivre 1974, p. 21.
40. Wigny 1992, pp. 245–246.
41. 1463–1494.
42. 1433–1499.
43. Blamont 1993, pp. 414, 819.
44. Lenoble 1957, p. 407.
45. 1493–1541.
46. Faivre 1974, pp. 15–16.
47. 1260–1327.
48. 1575–1624.
49. 1702–1782.
50. Faivre 1974, p. 23.
51. Faivre 1974, p. 33.
52. Gusdorf 1985, p. 141.
53. Porter 1997, p. 303.
54. Lichtenhaeler 1974.
55. 1781–1826.
56. 1783–1855.
57. 1813–1877.
58. Tsouyopoulos 1999, pp. 15–16.
59. Gusdorf 1984, p. 271.
60. 1768–1835.
61. 1755–1843.
62. Gusdorf 1984, p. 261.
63. Lohff 1990, p. 10.
64. 1776–1847.
65. Burdach 1848, p. 440.
66. Gusdorf 1984, p. 260.
67. Novalis, 1798, cited by Gusdorf 1984, pp. 279–280.
68. Lohff 1990, pp. 24–25.
69. Darwin 1794; Autenrieth 1801–1802, cited by Lohff 1990, p. 27.
70. Zimmermann 1803, cited by Lohff 1990, p. 27.
71. Schelling 1799, cited by Lohff 1990, p. 59.
72. Lichtenthaeler 1974.
73. Lohff 1990, pp. 205–207.
74. Müller 1825.
75. Helmholtz 1879, p. 10.
76. Hume 1748, para 23, 36.
77. Kant 1783, Vorwort IV 255.
78. Kant 1781.

Chapter 3

1. Goethe 1833.
2. Müller 1826a.
3. Koller 1958, p. 26.
4. Rothschuh 1953; Koller 1958, p. 29.

5. Koller 1958, p. 33.
6. 1771–1832.
7. Müller 1825, cited by Koller 1958, pp. 46–48.
8. Müller 1826a.
9. Müller 1826b.
10. 1817–1881.
11. Müller 1833.
12. Koller 1958, p. 232.
13. Müller 1826b, pp. III–IV.
14. Hagner 1992, pp. 31–32.
15. Tsouyopoulos 1992, pp. 65–66.
16. Hagner & Wahrig-Schmidt 1992, p. 7.
17. Müller 1825.
18. Müller 1825, cited by Tsouyopoulos 1992, p. 71.
19. Röschlaub 1816, cited by Tsouyopoulos1992, p. 70.
20. Tsouyopoulos 1992, p. 73.
21. Lohff 1990, p. 207.
22. Schelling 1806, cited by Lohff 1990.
23. Hörz, 1994, p. 57; Tsouyopoulos 1992, p. 77.
24. Lohff 1990, pp. 125–128.
25. Koller 1958, p. 72.
26. Müller 1826a, cited by Poggi 1992, pp. 198–199.
27. Koller 1958, pp. 72–81.
28. Eckermann 1836–1848.
29. Nietzsche 1883; Koller 1958, p. 81.
30. 1803–1873.
31. Turner, cited by Lenoir 1992, pp. 15, 44.
32. Turner, cited by Lenoir 1992, pp. 15–16.
33. Lenoir, 1992, pp. 22, 30.
34. Koenigsberger 1902, I, p. 48.

Chapter 4

1. Bichat 1800, cited by Pichot 1993, p. 526.
2. Cahan 1993b, pp. 95–96.
3. 1810–1882.
4. Holmes 1994, p. 9.
5. 1777–1859.
6. 1807–1893.
7. Bensaude-Vincent & Stengers, 1992, p. 271.
8. Holmes, 1994, p. 10.
9. Von Haller 1753.
10. Pichot 1993, pp. 348–349.
11. Grmek 1990, p. 138.
12. Rey 1997, p. 120.
13. Kant 1790.
14. 1765–1844; Kielmeyer 1793.
15. 1660–1734.
16. 1743–1794.
17. 1734–1806.
18. 1722–1776.
19. 1771–1802.
20. Bichat 1800.
21. Pichot 1993, p. 528.
22. Bernard 1865.
23. Debru 1983, p. 28.
24. 1850–1922; Hofmeister 1901; Debru 1982, pp. 92–93.
25. Müller 1833.

26. 1800–1882.
27. Holmes 1994, p. 12.
28. Mayer 1997, p. 275.
29. Canguilhem 1952, p. 95.
30. Helmholtz 1891, p. 9.
31. 1802–1870.
32. Helmholtz 1891, p. 10.
33. Helmholtz 1843, pp. 453–462.
34. Bensaude-Vincent & Stengers 1993, p. 272.
35. Helmholtz, cited by Koenigsberger 1902, I p. 59.
36. Helmholtz 1845, pp. 72–83.
38. Helmholtz 1846.
39. Olesko & Holmes 1993, pp. 60–61.
40. Helmholtz 1847a.
41. Koenigsberger 1902, I pp. 66–67; Kremer 1990.
42. Johannes 1902, cited by Koenigsberger 1902, I p. 66.
43. Kremer 1990, letters 22 & 33.
44. Eckermann 1836–1848, p. 261.

Chapter 5

1. Krüger 1994, p. 203.
2. Helmholtz 1847b.
3. Koenigsberger 1902, I p. 68.
4. Hörtz & Wollgast 1986.
5. Hörtz & Wollgast 1986, p. 25.
6. Koenigsberger 1902, I p. 243.
7. Rechenberg 1994, p. 10.
8. Hörtz & Wollgast 1986, p. 28.
9. du Bois-Reymond 1880, cited by Hörtz & Wollgast 1986, pp. 29, 45.
10. 1820–1895.
11. Koenigsberger 1902, I p. 69.
12. Bevilacqua 1993, p. 295.
13. Helmholtz 1847b, p. 1.
14. Helmholtz 1847b, pp. 1–4.
15. Bevilacqua 1993, p. 307.
16. Kant 1786, cited by Heidelberger 1993, p. 466.
17. Helmholtz 1847b, p. 1.
18. Heidelberger 1993, pp. 466–473.
19. Helmhotz 1847b, p. 14.
20. 1824–1907.
21. 1820–1872.
22. Rankine 1853, p. 106.
23. Bevilacqua 1993, p. 315.
24. Bevilacqua 1993, pp. 316–317.
25. Costabel 1997.
26. de Maricourt 1269.
27. Costabel 1997.
28. 1796–1832.
29. 1799–1864.
30. Rechenberg 1994, p. 58.
31. Helmholtz 1847b, p. 53.
32. Koenigsberger 1902, I p. 85.
33. Planck 1906, cited by Rechenberg 1994, pp. 58–59.
34. Helmholtz 1853a.
35. Heidelberger 1993, p. 483.
36. 1816–1895.

Intermezzo with Artists

1. Koenigsberger 1902, I pp. 94–95.
2. Koenigsberger 1902, I pp. 95–105.

Chapter 6

1. Rilke, cited by Arendt 1961, pp. 44, 285.
2. Helmholtz 1850b, 1851.
3. Piccolino 2003.
4. Rechenberg 1994, p. 72.
5. Helmholtz 1850a, pp. 276–277.
6. Weber 1846, cited by Olesko & Holmes 1993, pp. 80–81.
7. Helmholtz 1891.
8. Koenigsberger 1902, I p. 118.
9. Helmholtz 1850b.
10. Olesko & Holmes 1993, p. 88.
11. Helmholtz 1850b.
12. Marey 1873.
13. Helmholtz 1851.
14. Koenigsberger 1902, I pp. 129–130.
15. Koenigsberger 1902, I pp. 122–123.
16. Helmholtz 1867, pp. 228–234.
17. Helmholtz 1871, pp. 333–337.
18. Exner 1868.
19. Koenigsberger 1902, I pp. 133–134.
20. Prévost 1810, cited by Helmholtz 1856, I.
21. Cumming 1846.
22. Davson 1962, pp. 193–196.
23. Keeler 2004.
24. Helmholtz 1856, I p. 257.
25. Grüsser 1994.
26. Tuchman 1993, p. 33.
27. Rechenberg 1994, p. 84.
28. Helmholtz 1891, pp. 12–13.
29. Koenigsberger 1902, I p. 223.
30. Helmholtz 1856–1866.
31. Helmholtz 1855.
32. Helmholtz 1853a.

Chapter 7

1. Lacoste 1997, p. 13.
2. Eckermann 1836–1848, p. 393.
3. 1760–1817.
4. Lacoste 1997.
5. Goethe 1784b, cited by Fink 1991, p. 15.
6. 1707–1788.
7. 1769–1832 ; Lacoste 1997, pp. 175–176.
8. Lacoste 1997, p. 177.
9. Lacoste 1997, p. 180.
10. Eckermann 1836–1848, p. 274.
11. Lacoste 1997, p. 182.
12. 1741–1801.
13. Lacoste 1997, p. 43.
14. Goethe 1784a, cited by Fink 1991, p. 21.
15. Goethe 1830.

16. Eckermann 1836–1848, p. 616.
17. Darwin 1861.
18. 1707–1778.
19. Goethe 1790, cited by Fink 1991, p. 26.
20. Lacoste 1997, pp. 27–33.
21. Lacoste 1997, pp. 39–40.
22. Cassirer 1945.
23. Goethe 1790, p. 266.
24. Gusdorf 1985, p. 89.

Chapter 8

1. Goethe 1810, p. 80.
2. Goethe 1810, p. 80.
3. Sabra 1967.
4. Ballas 1997, p. 253.
5. Fink 1991, p. 32.
6. Goethe 1810.
7. Lacoste 1997, p. 93.
8. Gusdorf 1985, p. 93.
9. 1786–1889.
10. Ballas 1997, p. 34.
11. Chevreul 1839.
12. Ballas 1997, p. 108.
13. Chevreul 1832, para 65–82.
14. Roque et al. 1997, p. 137; Roque 1997, p. 16.
15. 1861–1925.
16. Ballas 1997, pp. 183–185.
17. Lacoste 1997, p. 126.
18. Goethe 1810, Introduction.
19. Helmholtz 1853a.
20. 1773–1829.
21. 1629–1695.
22. 1788–1827.
23. Ronan 1983.
24. Helmholtz 1852b.
25. Helmholtz 1856–1866, pp. 318–319.
26. Unger 1854, 1855; Helmholtz 1856–1866, pp. 356–357.
27. Helmholtz 1852b, pp. 600–601.
28. Helmholtz 1852b, p. 592.
29. Finger 1994, pp. 97–100.
30. Young 1802.
31. Turner 1994, pp. 96–98.
32. Helmholtz 1856–1866, pp. 367–408.
33. 1831–1879.
34. Turner 1994, pp. 99–104.
35. Helmholtz 1856–1866, p. 546.
36. Helmholtz 1868, pp. 325–326.
37. Mafei & Fiorentini 1995; Piccolino & Moriondo 2002.
38. Chevreul 1839.
39. Rood 1879.
40. 1859–1891.
41. 1863–1935.
42. Roque 1997, p. 313.
43. Ballas 1997, p. 43.
44. Helmholtz 1876, II pp. 95–135.

Chapter 9

1. Bernard 1850–1860, p. 145.
2. Koenigsberger 1902, I pp. 194–200.
3. Koenigsberger 1902, I p. 347.
4. Koenigsberger 1902, I p. 213.
5. Helmholtz 1855.
6. Koenigsberger 1902, I p. 243.
7. Helmholtz 1893, cited by Koenigsberger 1903, III pp. 118–119.
8. Helmholtz 1856–1866, p. 1000.
9. 1820–1900.
10. 1834–1918.
11. Turner 1994.
12. Helmholtz 1852b, pp. 591–609.
13. Helmholtz 1879, pp. 9–52.
14. Müller 1826a, pp. 44–55.
15. Leibniz 1704.
16. Carpenter 1842, cited by Gauchet 1992, p. 47.
17. Leibnitz 1704.
18. Helmholtz 1852b.
19. Turner 1994, pp. 10–31.
20. Turner 1993, p. 161.
21. Lotze 1852, pp. 197–206; Ribot 1879.
22. Heidelberger 1993, pp. 461–497.

Chapter 10

1. Goethe 1833.
2. Kirsten 1986, Letters 52, 54.
3. Koenigsberger 1902, I p. 267.
4. Helmholtz 1857, p. 121.
5. Helmholtz 1863.
6. Békésy 1964.
7. Helmholtz 1863, p. 2.
8. Helmholtz 1863, pp. 588–591.
9. Koenigsberger 1902, I pp. 23–29.
10. Cahan 1993b, Letter 6.
11. 1804–1849.
12. Koenigsberger 1902, I p. 33.
13. Kremer 1990, p. 195.
14. Koenigsberger 1903, II pp. 69–73.
15. Koenigsberger 1903, II p. 74.
16. Koenigsberger 1903, II p. 232.
17. Lavignac 1895, p. 9.
18. Koenigsberger 1903, II p. 232.
19. Koenigsberger 1903, III p. 2.
20. Werner & Irmscher 1992, p. 26.
21. 1830–1894.
22. Vogel 1993, p. 261.
23. 1756–1827; Chladni 1787.
24. Helmholtz 1863, p. 600.
25. 1768–1830.
26. Helmholtz 1863, pp. 55–56.
27. Helmholtz 1863.
28. 1822–1876; Corti 1851.
29. 1841–1897; Preyer 1876.
30. 1836–1921; Waldeyer 1872.

31. Vogel 1993, p. 280.
32. 1835–1924.
33. 1841–1922.
34. Hensen 1863b; Hasse 1867.
35. Hensen 1863a.
36. Helmholtz 1863, p. 125.

Chapter 11

1. Lévi-Strauss 1964, p. 26.
2. Helmholtz 1863, p. 369.
3. Helmholtz 1863, p. 274.
4. Helmholtz 1863, pp. 269–270.
5. Helmholtz 1863, pp. 315–316.
6. Helmholtz 1863, p. 253ff.
7. Tartini 1754.
8. Helmholtz 1863, pp. 315–316.
9. Helmholtz 1863, pp. 316–382.
10. Helmholtz 1863, pp. 277–283.
11. Helmholtz 1863, p. 379.
12. 1895–1963; cited by Fichet 1996, pp. 84–85.
13. Helmholtz 1863, pp. 385–386.
14. Helmholtz 1863, pp. 391–411.
15. Helmholtz 1863, p. 395.
16. Fetis 1835, I p. 126; Helmholtz 1863, p. 395.
17. Helmholtz 1863, p. 396.
18. 1561–1633.
19. 1560–1627.
20. 1517–1590.
21. 1483–1546.
22. Luther 1912–1921.
23. 1496–1544.
24. Helmholtz 1863, pp. 407–409.
25. Helmholtz 1863, pp. 412–416.
26. Helmholtz 1863, pp. 421–422.
27. 1645–1706.
28. Helmholtz 1863, pp. 505–507.
29. Helmholtz 1863, pp. 507–508.
30. Helmholtz 1863, p. 516.
31. 1794–1870.
32. Hörz 1994, p. 325.
33. Helmholtz 1863, p. 371.
34. Helmholtz 1847b: see chapter 5.
35. 1801–1887; Fechner 1876, cited by R. Bouveresse 1995, p. 23.
36. Helmholtz 1847b, p. 529.
37. Goethe, Faust.
38. Helmholtz 1847b, p. 590.

Chapter 12

1. Humboldt 1845–1862, cited by Duviols & Minguet 1994.
2. Mach 1866.
3. James 1890, I pp. 627–628.
4. Koenigsberger 1903, II p. 88.
5. 1849–1936.
6. Pavlov 1926, translated by Anrep 1960, p. 151.
7. Hörz & Wollgast 1986, p. 48.

8. Koenigsberger 1903, II p. 241.
9. Helmholtz 1892, II p. 337.
10. Schiemann 1997, pp. 375–381.
11. Helmholtz 1892.
12. Goethe 1810.
13. Kant 1790.
14. Luke 1964 (Goethe, Faust Part Two, lines 12104 ff).
15. Holl & Reschke 1999, pp. 12–15.
16. Humboldt 1847, cited by Gusdorf 1985, pp. 105–106.
17. Eckermann 1836–1848: December 11, 1826.
18. Lescourret 1999, p. 269.
19. Gusdorf 1985, pp. 141, 343.
20. Kümmel 1985, pp. 195–210.
21. Helmholtz 1850b.

Postface

1. Helmholtz 1892.
2. Hatfield 1993, p. 553.
3. Helmholtz 1876, II p. 135.
4. Todorov 1996, pp. 2–33.
5. 1671–1713.
6. Cassirer 1932.
7. Todorov 1996, p. 308.
8. Bouveresse 1995, I p. 95.
9. Merleau-Ponty 1945.
10. Cahan 1993a, pp. 561–562.
11. Cahan 1993a, p. 600.
12. Bouveresse J, 1995, I p. 95.

Bibliography

Adamov-Autrusseau, Jacqueline. 1972. L'Aufklärung. Le Romantisme. In *Histoire de la philosophie. Idées, doctrines. Les Lumières Le XVIIIe siècle*, ed. François Chatelet. Paris: Hachette.

Alexandre, Maxime. 1963. *Romantiques allemands (Introduction) Bibliothèque de la Pléiade*. Paris: Gallimard.

Arendt, Hannah. 1961. *Between Past and Future: Six Exercises in Political Thought*. New York: Viking Press.

Aromatico, Andrea. 1996. *Alchimia. L'oro della conoscenza*. Turin: Electa Gallimard.

Autenrieth, Johann Heinrich von. 1801–1802. *Handbuch der empirischen menschlichen Physiologie*. Tübingen: Heerbrandt.

Ballas, Giulia. 1997. *La couleur dans la peinture moderne. Théorie et pratique*. Paris: Biro.

von Békésy, Georg. 1964. Concerning the Pleasures of Observing, and the Mechanics of the Inner Ear. In *Nobel Lectures, Physiology or Medicine 1942–1962*. Amsterdam: Elsevier.

Bensaude-Vincent, Bernadette, and Isabelle Stengers. 1993. *Histoire de la chimie*. Paris: La Decouverte.

Bernard, Claude. 1865. *Introduction à l'étude de la médecine expérimentale*. Paris: Baillière. English translation: Henry Copley Greene. 1927. *An introduction to the study of experimental medicine*. New York: Macmillan.

Bernard, Claude. 1850–1860. *Cahier de notes 1850–1860*, ed. Mirko Drazen Grmek. 1965. Paris: Gallimard.

Besnier, Jean-Michel. 1993. *Histoire de la philosophie moderne et contemporaine. Figures et oeuvres*. Paris: Grasset.

Bevilacqua, Fabio. 1993. Helmholtz's "Ueber die Erhaltung der Kraft." The emergence of a theoretical physicist. In *Hermann von Helmholtz and the foundations of nineteenth-century science*, ed. David Cahan. Berkeley, Los Angeles: University of California Press, 291–333.

Bichat, Xavier. 1800. *Recherches physiologiques sur la vie et la mort*. Paris: Brosson, Gabon.

Blamont, Jacques. 1993. *Le chiffre et le songe. Histoire politique de la découverte*. Paris: Odile Jacob.

Bouquiaux, Laurence. 1995. Introduction. In *Discours de métaphysique, suivi de Monadologie*. Paris: Gallimard.

Bouveresse, Jacques. 1995. *Langage, perception et réalité: Vol. 1. La perception et le jugement*. Nîmes: Chambon.

Bouveresse, Renee. 1995. *Esthétique, psychologie et musique*. Paris: Vrin.

Bruno, Giordano. 1591. *De immenso et innumerabilibus.*

Burdach, Karl Friedrich. 1848. *Rückblick auf mein Leben (Blicke ins Leben),* vol. 4. Leipzig: Voss.

Cahan, David. 1993a. Helmholtz and the Civilizing Power of Science. In *Hermann von Helmholtz and the Foundations of Nineteenth-Century Science,* ed. David Cahan. Berkeley, Los Angeles: University of California Press, 559–601.

Cahan, David. 1993b. *Letters of Hermann von Helmholtz to His Parents: The Medical Education of a German Scientist 1837–1846.* Stuttgart: Franz Steiner Verlag.

Canguilhem, Georges. 1952. *La connaissance de la vie.* Paris: Hachette.

Carpenter, William B. 1842. *Principles of Human Physiology, with Their Chief Applications to Pathology, Hygiene, and Forensic Medicine.* London: Churchill.

Cassirer, Ernst. 1923–1929. *Philosophie der symbolischen Formen,* 3 vols.

Cassirer, Ernst. 1932. *Philosphie der Aufklärung.* Tübingen: Mohr. English translation: Fritz C.A. Koelln & James P. Pettegrave. 1951. *Philosophy of the Enlightenment.* Princeton: Princeton University Press.

Cassirer, Ernst. 1945. *Rousseau, Kant, Goethe.* English translation: James Gutmann, Paul Oskar Kristeller, and John Herman Randall. 1945. Princeton: Princeton University Press.

Chevreul, Michel-Eugène. 1832. Mémoire sur l'influence que deux couleurs peuvent avoir l'une sur l'autre quand on les voit simultanément. Mémoires de l'Académie des Sciences, Paris 11:447–520.

Chevreul, Michel-Eugène. 1839. *De la loi du contraste simultané des couleurs et de l'assortiment des objects colorés. Pitois-Levrault, Paris. English translation: Charles Martel 1854 The principles of harmony and contrast of colours, and their applications to the arts.* London: Longman.

Chladni, Ernst. 1787. *Entdeckungen über die Theorie des Klanges.* Leipzig: Weidmanns Erben und Reich.

Corti, Alfonso. 1851. Recherches sur l'organe de l'ouïe des mammifères. Zeitschrift für Wissenschaftliche Zoologie 3:1–106.

Costabel, Pierre. 1997. *Mouvement perpetuel. Encyclopaedia Universalis, Paris* 12: 796–797.

Cumming, William. 1846. On a Luminous Appearance of the Human Eye, and Its Application to the Detection of Disease of the Retina and Posterior Part of the Eye. Medico-Chirurgical Transactions 29:283–296.

d'Alembert, Jean-Baptiste Le Rond. 1751. *Discours préliminaire.*

Darwin, Charles. 1861. *The Origin of Species by Means of Natural Selection.* 3rd ed. London: Murray.

Darwin, Erasmus. 1794. *Zoonomia, or, The laws of organic life.* London: Johnson.

Davson, Hugh. 1962. *The Eye: Vol. 4. Visual Optics and the Optical Space Sense.* New York: Academic Press.

Debru, Claude. 1983. *L'esprit des protéines. Histoire et philosophie biochimiques.* Paris: Hermann.

Descartes, René. 1637. *Discours de la méthode.* Leiden: Jean Maire.

Desné, Roland. 1972. La philosophie française au XVIIIe siècle. In *Histoire de la philosophie. Idées, doctrines. Les Lumières. Le XVIIIe siècle,* ed. François Chatelet. Paris: Hachette.

du Bois-Reymond, Emil. 1880. Die sieben Welträtsel. In *Vorträge über Philosophie und Gesellschaf,* ed. Siegfried Wollgast. (1974) Hamburg: Meiner.

Duviols, Jean-Paul, and Charles Minguet. 1994. *Humboldt. Savant-citoyen du monde.* Paris: Découvertes Gallimard.

Eckermann, Johann Peter. 1836 (Vols. 1, 2), 1848 (Vol. 3). *Gespräche mit Goethe in den letzen Jahren seines Lebens 1823–1832*. English translation: John Oxenford. 1850. *Conversations of Goethe with Eckermann and Soret*. London: Smith, Elder.

Exner, Sigmund. 1868. Ueber die zu einer Gesichtswahrnehmung nöthige Zeit. Sitzungsberichte der kaiserlichen Akademie der Mathematisch-naturwissenschaftliche Classe 58:601–632.

Faivre, Antoine. 1974. La philosophie de la nature dans le romantisme allemande. In *Histoire de la philosophie: Vol. 3. Du xix siècle à nos jours*, ed. Yvon Belaval. Paris: Gallimard, 14–45.

Fechner, Gustav. 1876. *Vorschule der Aesthetik*. Leipzig: Breitkopf & Härtel.

Femmel, G., ed. *1958–1979 Corpus der Goethezeichnungen (7 vols.)*. Leipzig: Seemann.

Fétis, François-Joseph. 1835–1844. *Biographie universelle des musiciens et bibliographie générale de la musique* (10 vols.). Brussels: Leroux.

Fichet, Laurent. 1996. *Les théories scientifques de la musique. XIXe et XXe siècles*. Paris: Vrin.

Finger, Stanley. 1994. *Origins of Neuroscience. A History of Explorations into Brain Function*. Oxford: Oxford University Press.

Fink, Karl J. 1991. *Goethe's History of Science*. Cambridge: Cambridge University Press.

de Gandillac, Maurice. 1973. Giordano Bruno. In *Histoire de la philosophie: Vol. 2. Le siècle des Lumières, La révolution kantienne*, ed. Yvon Belaval. Paris: Gallimard.

Gauchet, Marcel. 1992. *L'inconscient cérébral*. Paris: Seuil.

Goethe, Johann Wolfgang. 1773. *Von deutscher Baukunst*.

Goethe, Johann Wolfgang. 1784a. Dem Menschen wie den Tieren ist ein Zwischenknochen der obern Kinnlade zuzuschreiben. In *Werke, Hamburger Ausgabe: Vol. 13. Naturwissenschaftliche Schriften I*.

Goethe, Johann Wolfgang. 1784b. *Ueber den Granit. Naturwissenschaftliche Schriften I*.

Goethe, Johann Wolfgang. 1790. *Versuch die Metamorphose der Pflanzen zu erklären*. Gotha: Ettinger. English translation: Douglas Miller, 2009. *The Metamorphosis of Plants*. Cambridge, MA: MIT Press.

Goethe, Johann Wolfgang. 1810. *Zur Farbenlehre Cotta, Tübingen. English translation: Charles L. Eastlake. 1840. Theory of Colours*. London: Murray. 1970 edition, MIT Press.

Goethe, Johann Wolfgang. 1830. Annalen, oder Tag- und Jahreshefte. In *Ausgabe letzter Hand*, vols. 31 and 32. Stuttgart: Cotta.

Goethe, Johann Wolfgang. 1833. *Maximen und Reflexionen*.

Gregory, Richard Langton. 1970. *The Intelligent Eye*. London: Weidenfeld.

Grmek, Mirko Drazen. 1990. *La première révolution biologique. Réflexions sur la physiologie et la médecine du XVIIe siècle*. Paris: Payot.

Grüsser, Otto-Joachim. 1994. *Hermann von Helmholtz and His Synthetic Approach to Science* (Lecture, Utrecht, 28 October 1994).

Gusdorf, Georges. 1984. *L'homme romantique*. Paris: Payot.

Gusdorf, Georges. 1985. *Le savoir romantique de la nature*. Paris: Payot.

Hagner, Michael. 1992. Sinnlichkeit und Sittlichkeit. Spinozas "grenzenlose Uneigennützigkeit" und Johannes Müllers Entwurf einer Sinnesphysiologie. In *Johannes Müller und die Philosophie*, ed. Michael Hagner and Bettina Wahrig-Schmidt. Berlin: Akademie-Verlag, 29–44.

Hagner, Michael, and Bettina Wahrig-Schmidt. 1992. Vorbemerkung. In *Johannnes Müller und die Philosophie*, ed. Michael Hagner and Bettina Wahrig-Schmidt. Berlin: Akademie-Verlag, 7–10.

Hasse, Carl. 1867. Die Schnecke der Vögel. Zeitschrift fur Wissenschaftliche Zoologie 17:56–104.

Hatfield, Gary. 1993. Helmholtz and Classicism. The Science of Aesthetics and the Aesthetics of Science. In *Hermann von Helmholtz and the foundations of nineteenth-century science*, ed. David Cahan. Berkeley, Los Angeles: University of California Press, 522–558.

Heidelberger, Michael. 1993. Force, Law, and Experiment. The Evolution of Helmholtz's Philosophy of Science. In *Hermann von Helmholtz and the Foundations of Nineteenth-Century Science*, ed. David Cahan. Berkeley, Los Angeles: University of California Press, 461–497.

Heine, Heinrich. 1836. *Die romantische Schule*. Hamburg: Hoffmann & Campe.

Helmholtz, Hermann. Note: Several of Helmholtz's papers were collected as *Populäre wissenschaftliche Vorträge* and *Vorträge und Reden* (1865, 1876, 1884). Brunswick: Vieweg. The reference is only cited below if this was the first publication of a given paper. Some of Helmholtz's scientific works were also collected posthumously as *Wissenschaftliche Abhandlungen* (1882, 1883, 1895). Leipzig: Barth.

Helmholtz, Hermann. 1843. Ueber das Wesen der Fäulniss und Gährung. *Müller's Archiv für Anatomie, Physiologie und wissenschaftliche Medicin*, 453–462.

Helmholtz, Hermann. 1845. Ueber den Stoffverbrauch bei der Muskelaktion. *Archiv für Anatomie, Physiologie und wissenschaftliche Medicin* 72–83.

Helmholtz, Hermann. 1846. Wärme, physiologisch. In *Encyklopädisches Handwörterbuch der medicinischen Wissenschaften*, vol. 35. Berlin: Weit, 523–567.

Helmholtz, Hermann. 1847a. Theorie der physiologischen Wärmeerscheinungen. In *Fortschritte der Physik im Jahre 1845*, vol. 1. Berlin: Reimer, 346–355.

Helmholtz, Hermann. 1847b. *Ueber die Erhaltung der Kraft. Eine physikalische Abhandlung, Berlin Physical Society 23 July 1847*. Berlin: Reimer.

Helmholtz, Hermann. 1850a. Messungen über den zeitlichen Verlauf der Zuckung animalischer Muskeln und die Forpflanzungsgeschwindigkeit der Reizung in den Nerven. *Archiv für Anatomie, Physiologie und wissenschaftliche Medicin* 276–364.

Helmholtz, Hermann. 1850b. Note sur la vitesse de propagation de l'agent nerveux dans les nerfs rachidiens. *Comptes rendus de l'Académie des Sciences, Paris* 30:204–206.

Helmholtz, Hermann. 1850c. Ueber die Fortpflanzungsgeschwindichkeit der Nervenreizung. *Archiv für Anatomie, Physiologie und wissenschaftliche Medicin* 71–73.

Helmholtz, Hermann. 1851. Deuxième note sur la vitesse de propagation de l'agent nerveux. *Comptes rendus de l'Académie des Sciences, Paris* 33:262–265.

Helmholtz, Hermann. 1852a. Messungen über Fortpflanzungsgeschwindigkeit der Reizung in den Nerven. Zweite Reihe. *Archiv für Anatomie, Physiologie und wissenschaftliche Medicin* 199–216.

Helmholtz, Hermann. 1852b. Ueber die Natur der menschlichen Sinnesempfindungen. *Königsberger naturwissenschaftliche Unterhaltungen* 3:1–20 (Habilitation lecture, 28 June 1852, Königsberg).

Helmholtz, Hermann. 1853a. *Ueber Goethes naturwissenschaftliche Arbeiten* (Lecture, 1853, Königsberg). Allgemeine Monatsschrift für Wissenschaft und Literatur 1853:383–398.

Helmholtz, Hermann. 1855. *Ueber das Sehen des Menschen (Lecture, 27 February 1855, Königsberg)*. Leipzig: Voss.

Helmholtz, Hermann. 1856, 1860, 1866. *Handbuch der physiologischen Optik*. In *Allgemeine Encyklopädie der Physik*, 3 vols. Leipzig: Voss.

Helmholtz, Hermann. 1865. *Ueber die physiologischen Ursachen der musikalischen Harmonie* (Lecture, Bonn). Vorträge und Reden 1:119–155.

Helmholtz, Hermann. 1863. *Die Lehre von den Tonempfindungen als physiologische Grundlage für die Theorie der Musik*. Brunswick: Vieweg. [Sixth edition 1913] English translation: Alexander J. Ellis. 1885. *On the Sensations of Tone as a Physiological Basis for the Theory of Music*. London: Longmans.

Helmholtz, Hermann. 1865. *Eis und Gletscher (Lectures, Heidelberg and Frankfurt). English translation: Edmund Atkinson 1897 Popular Lectures on Scientific Subjects.* New York: Appleton.

Helmholtz, Hermann. 1867. Versuche über die Fortpflanzungsgeschwindigkeit der Reizung in den motorischen Nerven des Menschen, welche Herr N. Baxt aus Petersburg im physiologischen Laboratorium zu Heidelberg ausgeführt hat. Monatsbericht der Deutschen Akademie der Wissenschaften zu Berlin 29 (April):228–234.

Helmholtz, Hermann. 1868. Die neueren Fortschritte in der Theorie des Sehens. Preussische Jahrbücher 21:149–171, 263–290, 403–430.

Helmholtz, Hermann. 1871. Uber die Zeit, welche nötig ist, damit ein Gesichtseindruck zum Bewusstsein kommt. Resultate einer von Herrn N. Baxt im Heidelberger Laboratorium ausgeführten Untersuchung. Monatsbericht der Deutschen Akademie der Wissenschaften zu Berlin 8 (June):333–337.

Helmholtz, Hermann. 1876. Optisches über Malerei (1871 bis 1873). In *Vorträge und Reden* 2:93–135. Brunswick: Vieweg. [2nd ed. 1896]

Helmholtz, Hermann. 1877a. *Das Denken in der Medicin* (Lecture, 2 August 1877). Berlin: Hirschwald. I In *Vorträge und Reden* 2:165–189. Brunswick: Vieweg.

Helmholtz, Hermann. 1877b, 1896b. *Ueber die akademische Freiheit in der deutschen Universitäten* (Rectoral lecture, Berlin, 15 October 1877). Universitätsprogramm Berlin.I In Vorträge und Reden 2:91–211. Brunswick: Vieweg.

Helmholtz, Hermann. 1879. *DieThatsachen in der Wahrnehmung.* Berlin: Hirschwald.

Helmholtz, Hermann. 1891. Tischrede bei der Feier des 70. Geburtstages In *Ansprachen und Reden, 2. November 1891. Hirschwald.* Berlin:, 46–59.

Helmholtz, Hermann. 1892. *Goethe's Vorahnungen kommender naturwissenschaftlicher Ideen* (Lecture to Goethe Society, Weimar, 11 June 1892). Deutsche Rundschau 72:115–132.

Helmholtz, Hermann. 1893. Ueber den Ursprung der richtigen Deutung unserer Sinneseindrücke. Zeitschrift für Psychologie und Physiologie der Sinnesorgane l. 7:81–96.

Hensen, Victor. 1863a. *Studien über das Gehörorgan der Decapoden.* Leipzig: Engelmann (originally in *Zeitschrift für Wissenschaftliche Zoologie* 13:319–412).

Hensen, Victor. 1863b. *Zeitschrift für Wissenschaftliche Zoologie* 13:681.

Herder, Johann Gottfried, ed. 1773. *Von deutscher Art und Kunst.* Hamburg: Bode.

Hindemith, Paul. 1937. *Unterweisung in Tonsatz.* Mainz: Schott.

Hofmeister, Franz. 1901. *Die chemische Organisation der Zelle.* Brunswick: Vieweg.

Holl, Frank, and Kai Reschke. 1999. Alles ist Wechselwirkung. In *Alexander von Humboldt. Netzwerke des Wissens.* Bonn: Kunst- und Ausstellungshalle der BRD, 12–15.

Holmes, Frederic. 1994. The Role of Johannes Müller in the Formation of Helmholtz's Physiological Career. In *Universalgenie Helmholtz*, ed. Lorenz Krüger. Berlin: Akademie-Verlag.

Hörz, Herbert. 1994. *Physiologie und Kultur in der zweiten Hälfte des 19. Jahrhunderts. Briefe an Hermann von Helmholtz.* Marburg: Basilisken-Presse.

Hörz, Herbert, and Siegfried Wollgast. 1986. Hermann von Helmholtz und Emil du Bois-Reymond. Wissenschaftsgeschichtliche Einordnung in die naturwissenschaftlichen und philosophischen Bewegungen ihrer Zeit. In *Dokumente einer Freundschaft. Briefwechsel zwischen Hermann von Helmholtz und Emil du Bois-Reymond (1846–1894)*, ed. Christa Kirsten. Berlin: Akademie-Verlag, 11–64.

von Humboldt, Alexander. 1845–1862. *Kosmos. Entwurf einer physischen Weltbeschreibung*, 5 vols. Stuttgart: Cotta.

Hume, David. 1748. *An Enquiry Concerning Human Understanding.*

Im Hof, Ulrich. 1993. *Das Europa der Aufklärung.* Munich: Beck.

James, William. 1890. *The Principles of Psychology. 2 vols.* New York: Holt.

Kant, Immanuel. 1781. *Critik der reinen Vernunft.* Riga: Hartknoch.

Kant, Immanuel. 1783. *Prolegomena zu einer jeden künftigen Metaphysik die als Wissenschaft wird auftreten können.* Riga: Hartknoch.

Kant, Immanuel. 1784. Beantwortung der Frage: Was ist Aufklärung? Berlinisches Monatschrift 4:481–494.

Kant, Immanuel. 1786. *Metaphysische Anfangsgründe der Naturwissenschaft.* Riga: Hartknoch.

Kant, Immanuel. 1790. *Critik der Urteilskraft.* Berlin: Lagarde & Friederich.

Kaschuba, Wolfgang. 1995. Deutsche Bürgerlichkeit nach 1800. Kultur als symbolische Praxis. In *Bürgertum im 19. Jahrhundert,* vol. 2, ed. Jürgen Kocka. Göttingen: Vandenhoeck und Ruprecht, 92–127.

Keeler, C. Richard. 2004. Babbage the Unfortunate. British Journal of Ophthalmology 88:730–732.

Kielmeyer, Carl Friedrich. 1793. *Ueber die Verhältniße der organischen Kräfte unter einander in der Reihe der verschieden Organisationen.* Stuttgart.

Kirsten, Christa, ed. 1986. *Dokumente einer Freundschaft. Briefwechsel zwischen Helmholtz und Emil du Bois-Reymond (1846–1894).* Berlin: Akademie-Verlag.

Koenigsberger, Leo. 1902 (Vol. I), 1903 (Vols. II & III). *Hermann von Helmholtz.* Brunswick: Vieweg. English translation: Francis A. Welby. 1906. Oxford: Clarendon Press.

Koller, Gottfried. 1958. *Das Leben des Biologen Johannes Müller (1801–1858).* Stuttgart: Wissenschaftliche Verlagsgesellschaft.

Kremer, Richard Lynn, ed. 1990. *Letters of Hermann von Helmholtz to His Wife, 1847–1859.* Stuttgart: Steiner.

Krüger, Lorenz. 1994a. Helmholtz über die Begreiflichkeit der Natur. In *Universalgenie Helmholtz,* ed. Lorenz Krüger. Berlin: Akademie-Verlag, 201–215.

Kümmel, Werner F. 1985. Alexander von Humboldt und die Medizin. In *Alexander von Humboldt. Leben und Werk,* ed. Wolfgang-Hagen Hein. Ingelheim: Boehringer, 195–210.

Kuhn, D., ed. 1954–1964. *Die Schriften zur Naturwissenschaft.* Vols. 9–10. Weimar: Böhlau.

Lacoste, Jean. 1997. *Goethe. Science et philosophie.* Paris: PUF.

Lavignac, Albert. 1895. *Le voyage artistique à Bayreuth.*

Leibniz, Gottfried Wilhelm von. 1704. *Neue Abhandlungen über den menschlichen Verstand.* English translation: *Alfred Gideon Langley 1896 New essays concerning human understanding.* New York, London: Macmillan.

Leibniz, Gottfried Wilhelm von. 1714. *Monadologie.* English translation: Robert Latta. 1898. *The Monadology.* Oxford: Clarendon Press.

Lenoble, Robert. 1957. Origines de la pensée scientifique moderne. In *Histoire de la science, Encyclopédie de la Pléiade,* ed. Maurice Daumas. Paris: Gallimard.

Lenoir, Timothy. 1992. Laboratories, Medicine, and Public Life in Germany 1830–1848. Ideological Roots of the Institutional Revolution. In *The Laboratory Revolution in Medicine,* ed. Andrew Cunningham and Perry Williams. Cambridge: Cambridge University Press, 14–71.

Lescourret, Marie-Anne. 1998. *Goethe. La fatalité poétique.* Paris: Flammarion.

Lévi-Strauss, Claude. 1964. *Le cru et le cuit.* Paris: Plon.

Lichtenthaeler, Charles. 1974. *Geschichte der Medizin. Die Reihenfolge ihrer Epochen-Bilder und die treibenden Kräfte ihrer Entwicklung. 2 vols.* Cologne-Lövenich: Deutscher Ärzte-Verlag.

Lohff, Brigitte. 1990. *Die Suche nach der Wissenschaftlichkeit der Physiologie in der Zeit der Romantik. Ein Beitrag zur Erkenntnisphilosophie der Medizin.* Stuttgart: Fischer.

Lotze, Hermann. 1852. *Medicinische Psychologie oder Physiologie der Seele.* Leipzig: Weidmann.

Ludwig, Carl Friedrich. 1858–1861. Lehrbuch der Physiologie des Menschen, 2nd ed., 2 vols. Heidelberg: Winter'sche Verlagshandlung.

Ludwig, Emil. 1920. *Goethe. Geschichte eines Menschen.* 3 vols. Stuttgart: Cotta.

Luke, David, ed. 1964. *Goethe, Selected Verse.* Harmondsworth: Penguin Classics.

Luther, Martin. 1912–1921. *D Martin Luthers Werke, Kritische Gesamtausgabe,* 6 vols. *Weimar: Böhlaus Nachfolger* 3:3815.

Mach, Ernst. 1866. *Einleitung in die Helmholtz'sche Musiktheorie. Populär für Musiker dargestellt.* Graz: Leuschner & Lubensky.

Maffei, Lamberto, and Adriana Fiorentini. 1995. *Arte e cervello.* Bologna: Zanichelli.

Marey, Etienne Jules. 1873. *La machine animale. Locomotion terrestre et aérienne.* Paris: Ballière.

Maricourt, Pierre de. 1269. *Epistola de magnete.*

Marquet, Jean-François. 1993. Schelling. In *La philosophie allemande de Kant à Heidegger,* ed. Dominique Folscheid. Paris: PUF.

Mayr, Ernst. 1997. *This Is Biology. The Science of the Living World.* Cambridge: Belknap Press of Harvard University Press.

Merleau-Ponty, Maurice. 1945. *Phénomenologie de la perception.* Paris: Gallimard.

Müller, Johannes. 1825. *Von dem Bedürfnis der Physiologie nach einer philosophischen Naturbetrachtung.* Bonn.

Müller, Johannes. 1826a. *Zur vergleichenden Physiologie des Gesichtssinnes des Menschen und der Thiere neben einen Versuch über die Bewegungen der Augen und über den menschlichen Blick.* Leipzig: Cnobloch.

Müller, Johannes. 1826b. *Uber die phantastischen Gesichtserscheinungen. Eine physiologische Untersuchung.* Koblenz: Hölscher.

Müller, Johannes. 1833. *Handbuch der Physiologie des Menschen für Vorlesungen.* Koblenz: Hölscher. English translation: William Baly. 1842. *Elements of Physiology.* London.

Nietzsche, Friedrich. 1883. *Also sprach Zarathustra.* Chemitz: Schmeitzner.

Olesko, Kathryn M., and Frederic L. Holmes. 1993. Experiment, Quantification, and Discovery. Helmholtz's Early Physiological Researches (1843–1850). In *Hermann von Helmholtz and the Foundations of Nineteenth-Century Science,* ed. David Cahan. Berkeley, Los Angeles: University of California Press, 50–108.

Pavlov, Ivan Petrovic. 1926. *English Translation: Gleb Vassilievitch von Anrep 1927 Conditioned Reflexes.* New York: Dover.

Piccolino, Marco. 2003. A "Lost Time" Between Science and Literature. The "Temps Perdu" from Hermann Helmholtz to Marcel Proust. Audiological Medicine I:261–270.

Piccolino, Marco, and Andrea Moriondo. 2002. Retina e visione. Elogio dell'imperfezione. Naturalmente 15/2:1–13; 15/4: 3–18.

Pichot, André. 1993. *Histoire de la notion de vie.* Paris: Gallimard.

Planck, Max. 1906. Helmholtz' Leistungen auf dem Gebiete der theoretische Physik. Allgemeine deutsche Biographie, Bayerische Akademie der Wissenschaften 51:470–472.

Poggi, Stefano. 1992. Goethe, Müller, Hering und das Problem der Empfindung. In *Johannes Müller und die Philosophie,* ed. Michael Hagner and Bettina Wahrig-Schmidt. Berlin: Akademie-Verlag, 191–206.

Porter, Roy. 1997. *The Greatest Benefit to Mankind. A Medical History of Humanity.* New York: Norton.

Prévost, Bénédict. 1810. Considerations sur le brillant des yeux du chat et de quelques autres animaux. *Bibliothèque britannique* 14:196–211.

Preyer, William Thierry. 1876. *Ueber die Grenzen der Tonwahrnehmung*. Jena: Dufft.

Rankine, William J. M. 1853. On the General Law of Transformation of Energy. Philosophical Magazine 4:106.

Rechenberg, Helmut. 1994. *Hermann von Helmholtz. Bilder seines Lebens und Wirkens*. Weinheim: VCH.

Rey, Roselyne. 1997. L'âme, le corps et le vivant. In *Histoire de la pensée médicale en Occident*, vol. 2, ed. Mirko D. Grmek. Paris: Seuil.

Ribot, Théodule-Armand. 1879. *La psychologie allemande contemporaine. Ecole expérimentale*. Paris: Ballière.

Ronan, Colin Alistair. 1983. *The Cambridge Illustrated History of the World's Science*. Cambridge: Cambridge University Press.

Rood, Ogden Nicholas. 1879. *Modem Chromatics with Application to Art and Industry*. New York: Appleton.

Roque, Georges. 1997. *Art et science de la couleur*. Nîmes: Chambon.

Roque, Georges, Bernard Bodo, and Françoise Viénot, eds. 1997. *Michel-Eugène Chevreul. Un savant, des couleurs*. Paris: Museum national d'histoire naturelle.

Röschlaub, Andreas. 1816. Versuch über die Methodik und Pseudomethodik in der klinischen Medizin. Neues Magazin für die klinische Medizin l: 3–99.

Rothschuh, Karl Eduard. 1953. *Geschichte der Physiologie*. Berlin: Springer.

Sabra, Abdelhamid I. 1967. *Theories of Light from Descartes to Newton*. London: Oldbourne.

Schelling, Friedrich Wilhelm Joseph. 1797. *Ideen zu einer Philosophie der Natur als Einleitung in das Studium dieser Wissenschaft. English translation: Errol E. Harris and Peter Heath. 1988. Ideas for a Philosophy of Nature: As Introduction to the Study of This Science*. Cambridge: Cambridge University Press.

Schelling, Friedrich Wilhelm Joseph. 1799. *Einleitung zu dem Entwurf eines Systems der Naturphilosophie*.

Schelling, Friedrich Wilhelm Joseph. 1806. *Aphorismen über die Naturphilosophie*.

Schiemann, Gregor. 1997. *Wahrheitsgewissheitsverlust: Hermann von Helmholtz' Mechanismus im Anbruch der Modeme. Eine Studie zum Ubergang von klassischer zu moderner Naturphilosophie*. Darmstadt: Wissenschaftliche Buchgesellschaft.

Seidengart, Jean. 1989. Giordano Bruno. Encyclopaedia Universalis, Paris 3:602–605.

Sherrington, Charles Scott. 1906. *The Integrative Action of the Nervous System*. New York: Scribner's.

Spinoza, Baruch (Benedict de). 1677. *Ethica ordine geometrico demonstrate*. Amsterdam. English translation: Robert Harvey Monro Elwes. 1883. *The Ethics, Parts I and II*. London: Bell.

Streidt, Gert, and Klaus Frahm. 1996. *Potsdam. Die Schlösser und Gärten der Hohenzollem*. Cologne: Könemann.

Tartini, Giuseppe. 1754. *Trattato di musica*. Padua.

Tilliette, Xavier. 1973. Schelling. In *Histoire de la philosophie. Encyclopédie de la Pléiade*, vol. 2, ed. Yvon Belaval. Paris: Gallimard.

Todorov, Tzvetan. 1996. *Introduction to JW Goethe, Ecrits sur l'art*. Paris: Flammarion.

Tsouyopoulos, Nelly. 1992. Schellings Naturphilosophie. Sünde oder Inspiration für den Reformer der Physiologie Johannes Müller? In *Johannes Müller und die Philosophie*, ed. Michael Hagner and Bettina Wahrig-Schmidt. Berlin: Akademie-Verlag, 65–83.

Tsouyopoulos, Nelly. 1999. La philosophie et la médicine romantiques. In *Histoire de la pensée médicale en Occident*, vol. 3, ed. Mirko D. Gremek. Paris: Seuil.

Tuchman, Arleen. 1993. Helmholtz and the German Medical Community. In *Hermann von Helmholtz and the Foundations of Nineteenth-Century Science*, ed. David Cahan. Berkeley, Los Angeles: University of California Press, 17–49.

Turner, R. Steven. 1993. Consensus and Controversy: Helmholtz on the Visual Perception of Space. In *Hermann von Helmholtz and the Foundations of Nineteenth-Century Science*, ed. David Cahan. Berkeley: University of California Press, 154–203.

Turner, R. Steven. 1994. *In the Eye's Mind. Vision and the Helmholtz-Hering Controversy*. Princeton: Princeton University Press.

Unger, Friedrich Wilhelm. 1854. *Disque chromharmonique pour servir à expliquer les règles de l'harmonie des couleurs*. Göttingen.

Unger, Friedrich Wilhelm. 1855. Esthétique des couleurs. *Comptes rendu de l'Académie des Sciences, Paris* 40:239–245.

Vogel, Stephan. 1993. Sensation of Tone, Perception of Sound, and Empiricism. In *Hermann von Helmholtz and the Foundations of Nineteenth-Century Science*, ed. David Cahan. Berkeley, Los Angeles: University of California Press, 259–287.

Von Haller, Albrecht. 1753. De partibus corporis humani sensibilibus et irritabilibus. *Commentarii Societatis Regiae Scientarium Gottingensis* 2:114–158. English translation: Owsei Temkin. 1936. A Dissertation on the Sensible and Irritable Parts of Animals. *Bulletin of the History of Medicine* 4:651–699.

Waldeyer, Wilhelm. 1872. Hörnerv und Schnecke. In *Handbuch der Lehre von den Geweben des Menschen und der Thiere*, vol. 2, ed. Salomon Stricker. Leipzig: Engelmann.

Weber, Eduard. 1846. Muskelbewegung. In *Handwörterbuch der Physiologie mit Rücksicht auf physiologische Pathologie*, vol. 3, part 2, ed. Rudolph Wagner. Brunswick: Vieweg, 1–122.

Werner, Petra, and Angelika Irmscher, eds. 1993. *Kunst und Liebe müssen sein. Briefe von Anna von Helmholtz an Cosima Wagner 1889 bis 1899*. Bayreuth: Druckhaus Bayreuth.

Wigny, Damien. 1992. *Sienne et le sud de la Toscane. Itinéraires, monuments, lectures*. Paris: Duculot.

Young, Thomas. 1802. On the Theory of Light and Colours (Bakerian lecture, Royal Society, 12 November 1801). *Philosophical Transactions of the Royal Society* 92:12–48.

Zimmermann, Ferdinand Joseph. 1803. *Philosophisch-medicinisches Wörterbuch zur Erleichterung des höherenmedicinischen Studiums*. Vienna: Camesina.